Wilding Ecologies, Walking-with Glacier

"Wilding Ecologies: Walking-with Glacier" is a poetic journey into the heart of our changing planet. The authors eloquent narrative weaves theory and pedagogy seamlessly, urging us to embrace a more interconnected, less anthropocentric worldview. As we walk alongside the melting glacier, the novella invites us to witness the entanglement of lives, human and non-human. Through prose, poetry, and captivating visuals, this book becomes a profound exploration of care and appreciation, a vital call to reawaken love and enchantment for our uncertain future.

—Michael A. Peters, *Distinguished Professor, Beijing Normal University, China*

"Wilding Ecologies: Walking-with Glacier" is a courageous and honest attempt of the authors to bear witness, to move beyond just thinking and talking about threatened wild places and landscapes. What is needed, they argue, is to live with and listen to those places too, like the Hardanger-jøkulen glacier in Norway. The group draws deep inspiration from visiting the elevated hut of the late ecophilosopher Arne Næss, who held that "the smaller we come to feel ourselves compared with the mountain, the nearer we come to participating in its greatness."

—Jan van Boeckel, *Professor Art & Sustainability, Hanze University of Applied Sciences, Groningen, The Netherlands*

"The wild pedagogues in this novella ask: What will it take to nurture caring, compassionate, and competent restorers of the earth? Can the idea of wild pedagogies provoke opportunities for reimagined relationships and experiences of enlarged human and more-than-human community? Walking with Glacier does not provide any definitive answers to these questions but it helps us slow down and contemplate in a deeply profound way that provides some guidance in where we might go next as a human species."

—Arjen Wals, *Professor Transformative Learning for Socio-ecological Sustainability, Wageningen University, The Netherlands*

"This book's wonderful set of detailed-rich stories from *wild pedagogues* on a trip to the Hardangerjøkulen glacier in Norway is as enjoyable to read, as it is informative and a call to radically transform environmental pedagogies.

It is an essential read for any environmental teacher who seeks to transform current (environmental) education and disrupt the Anthropocene saturated in Western ideologies that distance humans from the rest of Nature – the inquiry groundings of Wild Pedagogies."

—Greg Misiaszek, *Assistant Professor, Institute of Educational Theories, Beijing Normal University, China*

Karen Malone · Sean Blenkinsop ·
Bob Jickling · Marcus Morse

Wilding Ecologies, Walking-with Glacier

An Educational Novella

palgrave
macmillan

Karen Malone
School of Social Science,
Media, Film and Education
Swinburne University of Technology
Hawthorn, VIC, Australia

Bob Jickling
Faculty of Education
Lakehead University
Thunder Bay, Canada

Sean Blenkinsop
Faculty of Education
Simon Fraser University
Burnaby, BC, Canada

Marcus Morse
School of Education
University of Tasmania
Hobart, TAS, Australia

ISBN 978-3-031-63181-8 ISBN 978-3-031-63182-5 (eBook)
https://doi.org/10.1007/978-3-031-63182-5

Cover credit: Frozen in Paradox, Photographer Karen Malone

This Palgrave Macmillan imprint is published by the registered company Springer Nature Switzerland AG
The registered company address is: Gewerbestrasse 11, 6330 Cham, Switzerland

We dedicate this book to Aage Jensen, Marianne Presthus Heggen, as rousing wild pedagogues and organisers of the gathering in Finse, Norway, that inspired this novella. And to Hardangerjøkulen. Mountain, its boulders and rocks, the sun and clouds, the colour of blues and greys, the glacier, the streams, the cloudberries and the air caressing our faces as it passed us by. With much love and appreciation for the gift of being held by you all in this special place.

Aage Jensen

Telling cultural and medicinal tales of plants as they were being relayed in the author's nature journals, and for sharing his expansive knowing and doing of the Norwegian philosophy of *friluftsliv*, a simple and free life with nature, as we walked-with this extraordinary landscape.

Marianne Presthus Heggen

Sharing her life experience of being-with Glaciers, Marianne stands on the edge of Hardangerjøkulen.

Images: Photographers Karen Malone (Aage 1 & 2, Marianne 2) and Polly Knowlton Cockett (Marianne 1)

Acknowledgments

The authors would like to acknowledge Hardangerjøkulen Glacier as an active collaborator and source of inspiration in the writing of this book. We understand that this beautiful dome of ice is threatened by climate change.

We also acknowledge all those who have lived in this landscape, with the mountains and Glacier. We believe that within these cultures there are histories, ideas, memories, and myths that do not align with the dominant narrative that has given us anthropogenic climate change. We believe these disruptive notions might become elements for telling different stories.

Hardangerjøkulen Glacier, 35mm Film Image, Photographer Karen Malone

Contents

Notes on Contributors

Lee Beavington is a poet, learning strategist and interdisciplinary instructor at Kwantlen Polytechnic University, and has taught graduate students at Simon Fraser University in nature-based education. He coordinates KPU Wild Spaces, which focuses on place-based learning in higher education, and serves on the Climate+ Challenge instructor team. More at www.wildethic.com.

Chris Beeman, Ph.D. is a Professor in the Faculty of Education at Brandon University in Manitoba, Canada. His earned doctorate is from Queen's University at Kingston. He specialises in understanding the philosophical underpinnings of ecological learning, in part through an exploration of Indigenous theories of knowledge. His research explores ontologies of Indigeneity and environmental philosophy through long-term interactions with the more-than-human world.

Joshua David Bennett is an Ecosocial Educator, Outdoor guide, and host of the *Transnatural Perspectives Podcast*. With a Master's in Outdoor Education from the University of Marburg (Germany), Josh explores the landscapes where society and environment intersect. Based in Oslo Norway, Josh can be found guiding bike tours or finishing a Master's in Social Work, with a focus on Nature Based Therapy. Listen to the poem *Midnight Sun* included on his latest album "Songs at the end of the world" https://botik.bandcamp.com/track/midnight-sun-3

Sean Blenkinsop is a Professor in the faculty of education at Simon Fraser University, Vancouver, Canada. He undertook his doctoral work in philosophy of education at Harvard University with a focus on existentialism, education, dialogue, and the environment. His current research explores teacher education, school change, nature as co-teacher, and the challenges of justice and the environmental crisis in a rapidly changing world. Sean has published more than 100 articles and chapters. His most recent books are: *Wild Pedagogies: Touchstones for Re-Negotiating Education and the Environment in the Anthropocene* published by Palgrave-McMillan in 2018 and *Ecoportraiture: The Art of Research when Nature Matters* published by Peter Lang in 2022.

Polly Knowlton Cockett is involved in ecological literacy, sense of place studies, urban native biodiversity conservation, and socioecological cartography. Currently, she works as an Instructor in Teacher Education at the University of Calgary, Canada, and as President of Grassroutes Ethnoecological Association, she is engaged in creative community place-making.

Dr. Michael Dé Danann Datura (M.A., Ph.D., Simon Fraser University) is a humanities teacher, an educational researcher, and occasionally a poet. He is Principal of Cortes Island School, Surrealist, Microphone Controller. His research draws on lyric philosophy and the hermeneutic tradition to examine place-based education, Indigenous ways of knowing, and radical politics in the context of settler colonialism and ecological emergency. His work aims to cultivate an ecosophic disposition to recognise the experience of wisdom traditions, contemplative practices, and critical reflexivity in a more-than-human world. He also likes long walks in the forest. He is a member of the Academic Council for The Centre for Imagination in Research, Culture & Education (CIRCE: www.circe-sfu.ca)

Marianne Presthus Heggen is a Professor at Western Norway University of Applied Sciences in Bergen, Norway. Trained as an ecologist, her teaching and research spans around pedagogy of outdoor activities in early childhood. She tries to keep her research close to children and children's contribution to their own education, particularly through children's curiosity.

Per Ingvar Haukeland Professor in ecophilosophy and ecopedagogy, University of South-East Norway, teaches at the friluftsliv (outdoor life) education and heads the research group, Friluftsliv, Society and Sustainability. He has collaborated with Arne Naess over the last 20 years of Naess' life, including the books *Life's philosophy* (1998, translated in 2003) and *Dyp Glede: Inn i dypøkologien* (2008; Deep Joy: Into the depths of deep ecology; in translation). He is the co-founder of the Alliance for Wild Ethics, and heads the foundation *Tvergastein*: Arne Naess' hut. His current research interest is in eco-animism, indigeneity, and Sami/Norse nature-cultures.

Bob Jickling Professor Emeritus at Lakehead University, has interests in environmental education and philosophy. His current research attempts to find openings for radical re-visioning of education. His most recent books include *Wild Pedagogies: Touchstones for Re-Negotiating Education and the Environment in the Anthropocene* and *Environmental ethics: A sourcebook for educators*. As a long-time wilderness traveller, much of his inspiration is derived from the landscape of his home in Canada's Yukon.

Erika Kazi has a Master's in Education: Sustainability, Creativity, and Innovation from Cape Breton University and conducted research for the Social Science and Humanities Research Council of Canada. Erika continues to foster her research interests in sustainability, mental health, immersive recreation, and climate action as a steward for her community and industry. She is young and would like not to die from climate change because we're all too stupid and selfish to do anything about it. She lives in Salt Lake City with her rabbit and her dog and enjoys skiing and snowboarding, gardening, hiking, and reading.

Estella Carolye Kuchta is the author of *Finding the Daydreamer* (Elm Books). She teaches composition, literature, and research writing at Langara College in Vancouver, Canada. She holds a BFA in Creative Writing and an MA in Literature and is pursuing a Ph.D. in Ecological Education from Simon Fraser University.

Dr. Karen Malone is a Professor of Human Geography, Environmental and Childhood Studies in the Department of Education at Swinburne University of Technology. She is an international author, researcher, educator and photographer who studies human–environment interactions and the impacts of the ecological crisis including climate change, militarised radiation, and biodiversity loss on Earthly relations. In her teaching and research, she applies ecofeminist, posthuman, and Indigenous theoretical perspectives. Karen has published 12 books and over

100 articles. Her most recent books include *Children in the Anthropocene*, *Theorising Posthuman Childhood Studies* and *Reimagining Sustainability in Precarious Time*. You can find more about her research at: www.childrenintheanthropocene.com and her ORCID profile https://orcid.org/0000-0002-9904-7081

Deb Matlock is committed to nurturing connections between people, animals, earth, and spirit. She has spent 25 years working as an environmental educator and naturalist. Deb offers nature connection practitioner training, workshops, and retreats through her business, Wild Rhythms. She is completing a Ph.D. project on integrating interspecies communication into environmental education.

Marcus Morse is Associate Professor of Environmental, Outdoor and HPE at the University of Tasmania, Australia. He grew up in Tasmania, where he spent time on the island's rivers, coastlines, and mountains that inspired a love of being outdoors. Marcus' teaching is focused on place-based, attentive, and relational outdoor and environmental education. His research interests include environmental education, community engagement projects, place, river experience, and wild pedagogies.

Stacy Boe Miller is a poet and non-fiction writer living in the Pacific Northwest. Her work can be found or is forthcoming in Terrain.org, Copper Nickel, The Sun, and other journals. She is the current Poet Laureate of Moscow, Idaho.

Dr. Brant G. Miller is a Professor of Science Education in the Department of Curriculum and Instruction at the University of Idaho. Previously he was an eighth-grade science teacher in Western South Dakota.

Rebekka Sæter is a movement-based artist and outdoor educator from Oslo, Norway. She is the Artistic Director of the interdisciplinary art project "ghosting Glacier" and is the current head of the Norwegian BioArt Arena in Ås, Norway.

Maja Zimmermann is from Edmonton, Alberta, Canada and lives near Sogndal in western Norway where she and her partner are developing a smallholding. She has worked as a teacher and an outdoor educator, instructor and guide for various age groups and subjects, most recently teaching friluftsliv and nature guiding as programme coordinator of the bachelor programme in nature-based tourism at Western Norway University of Applied Sciences. She has a special passion for paddling, and enjoys exploring and relating with her home place through gardening, foraging, climbing and skiing.

List of Figures

CHAPTER 1

Walking-with Glacier

Drawn to the Glacier

This is a story book. In fact, it is a set of stories within a story book. The larger story—the container for the smaller episodes—is framed by a group of wild educators who gathered in the mountains of Norway. We, the gathering participants, were there out of a shared concern for an uncertain future for our human-selves and more-than-human neighbours. And as it turns out, the most compelling of these neighbours was the glacier Hardangerjøkulen and its middle arm, Midtdalsbreen. We walked up to them, touched them, witnessed their melting, wept for their passing, and, in some small ways, were awakened to their story. We talked, mused, and wrote—journal entries, poems, fragments of thought—and we made some photographs. What follows are the stories within.

We were based at Finse, a train stop at the high point along the Oslo to Bergen line. The gathering was one of a series of colloquia exploring ideas about what we call *Wild Pedagogies*. As wild pedagogues we all shared a view that as humans we cannot continue as we are. Effective responses to the crises of our times will need to be less anthropocentric, less hierarchical, and more just and equitable. We felt education must be a part of any response that requires such a fundamental rethinking of ideas and practices and, in turn, change (Fettes & Blenkinsop, 2023; Jickling et al., 2018a). And we shared a concern that our globalised world has knitted

K. Malone et al., *Wilding Ecologies, Walking-with Glacier*, https://doi.org/10.1007/978-3-031-63182-5_1

values, behaviours, and assumptions that seem near impossible to dislodge (Jensen et al., 2022).

The overarching idea of this series of colloquia has been to combine travel embedded in a landscape with thought-provoking seminars for participants to share ideas. We also shared a sense that you cannot just think and talk about places and landscapes and expect to be effective; we must live with and listen to those places too. It is through an ongoing series of immersive travelling dialogues that the ideas of Wild Pedagogies evolved, expanded, and, as it turns out, is how we arrived in Finse (Jensen et al., 2022).

In keeping with Norwegian traditions of rambling in a landscape, our mode of travel was walking—less structured at first. As we moved about, walking and thinking, our movements transformed into an everyday pedagogical practice. Then, as we began to listen to the call of the land, our feet and hearts drew us evermore towards the glacier and its proffered invitation. Our wandering research, it seemed, was being shaped by the landscape itself; and, in this way, the place itself emerged as a co-researcher in this project. (Blenkinsop, 2018; Blenkinsop et al., 2022).

Ours is a storied research method and we call the container for these stories—this book—a novella. In the spirit of the original novella—the *Decameron* (Boccaccio, n.d.)—these stories tell various versions of the news of our days together, inspired by key events and our sharing bodily experiences. Within the novella we have mapped this series of stories; they tell of our witnessing and weeping. And after a walk to Arne Næss' place called Tvergastein, we tell stories of awakening and work to be done.

Throughout this novella, we posit a need for loosening control of the social practices and embedded assumptions that tend to bend us back in the direction of the status quo, loosening those things that limit our curiosities and imaginations and compel us, often unwittingly, to continue thinking and being in the same ways that we have always done. Our response in Finse, and in earlier Wild Pedagogies gatherings, has been a wilding of the ecologies of our practices and, by extension, of ourselves as human beings and becomings.

Ecology, often associated with the science of ecosystems, has helped us to understand the importance of interdependent relationships within ecosystems. When, for example, wolves are removed from an ecosystem, that ecosystem becomes diminished. Yet, when the wolf is re-introduced, we appreciate that the whole ecosystem can, again, begin to flourish. Still, in this work we lean towards a more expansive vision of ecology stemming from its etymological origins in the Greek word *oikos*, often translated as meaning the *household*. Wilding ecologies in this novella is concerned with enriching and expanding the households of ecologies—households that include layers of self, cultures, and communities that can include Glacier, Spider, and other more-than-humans.

For example, Robert Bringhurst writes about an ecology of languages, suggesting that when we remove languages from the world, the entire ecosystem of languages, and more, is weakened. "When you kill a language off—even a language with only a single speaker—you make the entire planet less intelligent, less articulate, less capable—and decidedly less beautiful—than it was" (2002, p. 11). We agree. And we are inspired and impelled in this way toward a necessary re-wilding of ecologies more broadly.

Through the creative curation of the tales that follow we attempt to also expand ecosystems of knowledge, rationality, and ontology. And we aim to enrich these ecosystems with space for feelings, faint perceptions, gestalts, narrative expressions, and glaciers to flourish. There may be times where you, the reader, feel cast adrift by the variety of expressions—yet we hope that you might accept and enjoy this as part of a wild novella experience.

In the balance of this opening chapter, we provide additional background. We tell some of what we know about glacier lives—although Glacier tells us much more as the stories progress. We reflect on the experience of walking with glaciers, walking with the novella, walking with Wild Pedagogies, and throughout, we reflect on these experiences as wilding ecologies. And we reflect on how the work we are doing might change us, our communities, and trouble some aspects of our own cultures. We also begin to sprinkle about the first bits of news from our rambles.

Glacier Lives

The tale of glaciers commences with the ethereal dance of snowflakes. Producing glacial ice requires both snow and cold temperatures over unyielding winters (Jackson, 2019). In the life of our planet there have been many periods[1] where glaciers rebirthed and grew, moulding and reshaping the very essence of the world around them. And there were epochs when they relinquished their icy grip, melting into mists, and leaving behind indelible signs of their passing. Presently we find ourselves in an interglacial respite, a time of warmth conducive to the flourishing of mammals, nestled between the throes of ice ages. The last glacial period, or ice age, occurred between 120,000 and 11,500 years ago, while the most recent and albeit smaller, episode occurred between about 1300 and 1850 CE.[2] Oscillations in glacier formation are largely associated with the solar radiation penetrating the Earth's atmosphere. In contemporary times the rise of entrapped greenhouse gases within have kindled a sharp rise in global temperatures, triggering rapid thaws events that have ushered many glaciers into dramatic retreat.

Glaciers are giant rivers of ice that move across roughly a tenth of Earth's surface. Once, more than 400,000 of these icy titans veiled the planet's visage. Continental glaciers, or ice sheets, comprised the bulk of this frozen domain. While most glaciers cling to polar realms, outliers like our companion Hardangerjøkulen, find solace in mountainous enclaves, gracing every continent save Australia. These are guardians of the world's freshwater reserve. With the rhythm of seasons and warming breath of time, they birth meltwaters that cascade into lakes, rivers, and countless tributaries. In the intricate tapestry of temperate lands and mountainous horizons, this meltwater weaves the very fabric of existence, quenching the thirst of flora, fauna, and humanity alike.

Fig. 1.1 "Hardangerjøkulen Glacier #1" 35 mm Film image: Photographer Karen Malone

Fig. 1.2 "Hardangerjøkulen Glacier #2" 35 mm Film image: Photographer Karen Malone

Melting events produce profound alterations in the Earth's cryosphere—the domains where water exists in its solid form. The frozen weaving of glaciers, sea ice, river ice, ice caps, ice sheets, snow cover, and permafrost—reflects the sinews of Earth's climatic histories. The cryosphere influences the delicate dance between terrestrial moisture, atmospheric warmth, celestial clouds, tears of rain, and even the ebbs and flows of ocean currents. The ballet of melting glaciers, consequently, impels the inexorable ascent of sea levels—a disconcerting harbinger of slow onset climate change, as identified by the Intergovernmental Panel on Climate Change (IPCC). The arctic regions play a critical and potentially exponential role in global warming, and current data exposes a sombre narrative—glaciers spanning the globe, are shrinking, or disappearing altogether with glacier melting accounting for the biggest impacts on sea level rise.[3]

What transpires when a glacier disappears? Even as the essence of being a glacier is perpetual motion—a ceaseless metamorphosis between the solid and the fluid—what remains after millennia of existence? Jackson (2019) muses, "Glaciers are not mere monoliths of ice; they are labyrinths of tunnels, channels, and a myriad of other enigmatic features. As the glacier evolves, so too does its inner geography" (p. 87). Just as we do, when we stand with the glacier and witness a perpetual process of transformation.

In these precarious times, we are aware that what the future may become is constantly being made and remade each day. What is attainable, what is imagined, what might become? Glaciers and humanity are bound by entwined destinies. Throughout this novella, we endeavour to convey the awe, hope, and sorrow that encapsulate an enduring saga of renewal. We embark upon a journey of listening and speaking, sharing our kinship with the glacier—bonds forged in the belief that all things pulse with life, that glaciers are animate beings. As humans we stand witness to their vitality. In this text, we unravel stories that defy confinement, where boundaries blur, and where interdependence flourishes. In this novella, we loosen control over how we tell our stories, how we represent experiences in the world, and how we participate in an academic life.

Climate change, as an ecological call to arms, stands at the heart of the present human quandary. How do we confront the spectre of catastrophe? Fear and grief are our companions. The politics of mourning and loss, birthed from the crucible of climate upheaval, beckon us to make choices about what and who must be heard and who merits our emotional embrace. We are compelled to recalibrate our focus where need be (Muir et al., 2020). Such distinctions are pivotal in the theatre of place-politics, for they carve the demarcation lines between entities we acknowledge and those we deem unworthy of our attention. This disconnection bears a perilous edge, encouraging us to avert our gaze. It is a vision of post-modernity steeped in disenchantment. This sensation, often coined the Anthropocene, wherein we bear witness to the inexorable grip of death all around, can be profoundly disheartening.

But in response, Jane Bennett (2001, p. 3) presents us with a question:

> ...not whether disenchantment is a regrettable or a progressive historical development. It is, rather, whether the very characterization of the world as disenchanted ignores and the discourages affective attachment to the world.

As Jackson (2019) notes, it is dangerous to characterise the loss of glaciers as inevitable, their ruin *a fait accompli*; it is dangerous to assume that glaciers will be gone and therefore we need to acknowledge their loss and move on to the planned futures without them:

> Increasing yearly to the cacophonous discourse surrounding climate change, first person accounts of glacier retreat or disappearance in local landscapes underwrite both a glacier-ruins narrative and the climate change narrativized world, whereby more and more witnesses interconnect and substantiate the sheer ubiquitousness of the phenomenon. It is of little surprise that the arts, and more broadly the humanities, construct a world characterized, and, to some degree actualized, by the inevitability of pending glacier loss. (Jackson 2015, p. 488)

Feminist new materialist scholar Stacy Alaimo (2016), recognising the disjuncture of scale at the core of the climate crisis, suggests universalising frames have failed to advance understanding or required action. Situated (experiential) knowledge, she asserts, is needed—a trans-corporeal reality of human and nonhuman entanglements. What if our desire, instead, was to spend time with Glacier to restore love and enchantment, wonder, and awe? An enchantment that could inspire and render deep attachment in a more-than-human-world.

This novella is the container for storying our encounters. It is the concept for framing our experiences as wonder and enchantment rather than the grief of eco-nostalgia or lost ecological possibilities. In this work we map a series of stories tied with key events that have influenced our sharing of bodies, of walking, and of thinking.

It is our journey in the company of a particular community.

And these paths lead us further up the mountain, closer to and, at times, further away from Glacier.

Walking-with Glacier

Walking-with Glacier is walking to notice, to attune with, and to be affected by sensorial knowing (Malone & Moore, 2019). Bodies sweat—and melt—and do heavy lifting with and through the unknowing; "walking is embodied because it is immediate, tangible, and foregrounds the bodily experience of moving" (Springgay & Truman, 2017, p. 30). Walking-with Glacier "reminds us of our ancient chimeric entanglements" (Tsing et al., 2017, M2) and the evils wrought by Modern man (sic) and help us to notice landscapes of entanglement, bodies with other bodies, time with other times (Tsing et al., 2017).

The *Climate in Norway 2100* report, released by the Norwegian Centre for Climate Services (NCCS), projects that by the end of the century large glaciers in Norway will lose a third of their area and volume. Smaller glaciers, except at the very highest altitudes, and are expected to disappear entirely:

> For large glaciers a reduction of up to 1/3 of the area and volume they have today is expected, such that the remaining glaciers will be significantly smaller and will only be found at higher altitudes. The largest changes in glacier area and volume are expected to occur after the middle of this century. (Hanssen-Bauer et al., 2017. p. 33)

This wild ecological novella walks theory and pedagogy in the shadow of Hardangerjøkulen, Norway's sixth largest mainland glacier, near Finse. Walking as a pedagogical practice of everyday life (Sandlin et al., 2011) is a walking-with openness; it is a practice of circulating within sites of rocks, ice, ancient stories, vanished reindeer, forgotten peoples, water, sky, spiders, and plants all mangled together. "As we walk, we are 'in' the world, integrating body and space coextensively" (Springgay & Truman, 2017, p. 30). Walking-with Glacier is walking with an ancient ice flow riveted across time in a beloved landscape; it is an assemblage of more-than-human entities encountering and creating a shared place.

As we take the journey of this novella, and travel at scale beyond our comprehension, we hope you will share in our walking-with Glacier. As we walk heel to toe to earth with glimpses of the light shimmering over the ice sheet, our bodies simultaneously stretch across an immanence of the world and a transcendence.

We came to the mountains of Norway with a shared fear of losing something; perhaps something we never had, something we thought we loved, and something from which a sense of seeming "eco-nostalgia" emerges (Braidotti, 2019). To walk-with and witness Glacier's meltingness is to ask: What is being lost? What is being uncovered? What is being released? And how are beings in this place changed?

The temporal scale of climate change, as a critical manifestation of the Anthropocene, helps us to make sense of what different durations might mean in relation to one another and challenges us to walk differently, not just as mere spatial dimensions. Expressions of the Anthropocene have exposed human limitations of linear time. As we walk, we are always fixed in the in-between. We are not used to thinking of time dimensions exceeding our own lifetimes or of those demanding radical and immediate transformation (Macfarlane, 2019). We are held in the present, in-between what was and what could have been, and what is and what may be.

At one juncture, as we walked through this tampered landscape, we paused to be present with a large piece of bedrock. The place spoke of its tremendous power and longevity through this rock. It spoke to us through a connection to an Earth under constant construction, constant change, constant evolution. Being present, we felt layers in the rock as we ran our fingers across its sensuous ridges. We glimpsed into deep time, into geological time, into billions of years traversed under the sweep of our fingers.

These large rocks are scattered across the landscape as evidence of the retreating glacier. Transported by the ice, such glacial erratics are often made up of rock types different from the rocks found supporting them in the places where they now rest. They are a record of Glacier's travels and their journey together. And they remind us to always consider central questions of the posthuman—Where am I? Who am I here with?

The world is constantly being made and remade through sensorial moments. During times of walking, when we might pause and pay attention, to listen to the winds, and to touch the moraines, we are drawn into ancestral histories. Rock and land shaped by time and Gaia and the long unbroken chain of creation.

Fig. 1.3 "Deep Time bedrock, megablock" 35 mm Film image: Photographer Karen Malone

Fig. 1.4 "Glacier erratics" 35 mm Film image: Photographer Karen Malone

As a Glacier silently creeps across the landscape, advancing and retreating it drags rock, sediment, and ice over the bedrock beneath.

Littering

As I flow over the land, I pick up rocks in my sole deep below the surface where Glacial ice walks with the bedrock. Out of sight, shuffling over the rock surface, I remove material from my path. I make it smoother, detach layers of rocky debris, or roughen the bedrock to create cracks and openings. These openings become spaces deep beneath the cold and ice that give birth to new possibilities. Reshaping. Grinding. Scraping. Scratching. Sculpting. This abrasive agency gives liveliness to my flow; I am shaping the landscape below and beyond icy edges.

Huge chunks of bedrock are plucked from these cracks and are carried on my ice, tumbling over and under me and the meltwater. Finally, I leave them hundreds of kilometres from their birthplace, as rock litter. These glacier erratics record the story of my travels. As humans run hands across their surface and look deeply into the layers and patterns etched in the rock, they come to know they are reaching into different times, beyond them-selves. Touching them reveals complex patterns and flows outside of comprehension, at a time scale of geologic events vastly greater than any time scale of Anthropos—beyond human, before human. These glacial erratics reveal deep time stories; they are gifts that can reflect on the temporality of time and scale. It is almost impossible for humans to imagine the scale of something as vast as geological time. Deep time.

Deep time is one of geology's great gifts to the world's capacity for thinking new ideas. Human exceptionality fades, anthropocentrism dims as humans walk in the shadow of these huge glacial rocks sitting like giant gravestones on a diminishing landscape as they head towards my receding edges

Knight writes “Glaciers are one of nature’s most efficient landscape architects, constantly reshaping their environments through the processes of erosion, entrainment, transportation, and deposition” (1999, p. 26).

We paused to touch, to feel, to show our gratitude to the life of Glacier, and its relations with the emergence of Earth, its place in other moments of time, and the way it holds open the living present, and immanent field of possibilities, waiting beyond the horizon (Abram, 2017).

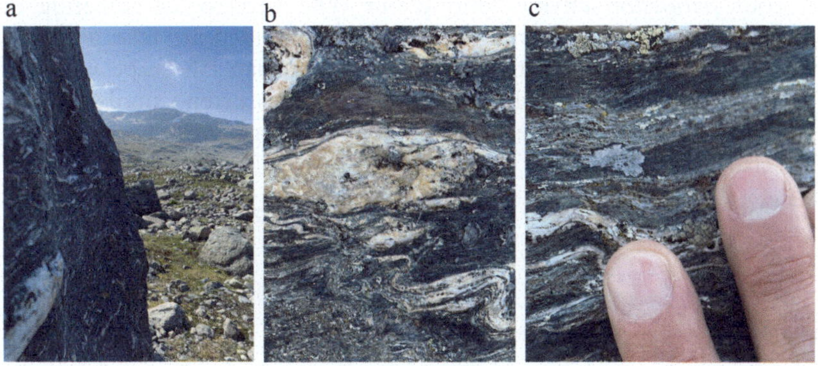

Fig. 1.5 “Touch is world-making”—Digital Images Photographer Karen Malone

Touch is world-making; it is a thought that resonates with relational ontologies wherein being and becoming moves in, out, and through myriad relationships. In this, we can look to Karen Barad’s account of the see-touching made possible by “scanning tunnelling microscopes” that bring us close to the particularities of place (Barad 2007, p. 309). To touch and to be touched. To be related to and with. We begin to understand the world, literally and as a possibility, in the meeting of knowing and being. We learn through being in the world rather than separate from it. In touching, we are engaging directly with the time passage of Earth’s emergence. We start to question who and what has this rock-being become? Who and what is being touched through this encounter? If touch is an experience in which boundaries of self and other blur, is it ever possible to touch without being touched by the other and their needs? Touch “ramifies and shapes accountability” (Haraway 2008, p. 36) just as relating engages both pleasure and obligation. Touching is world-making; the rich nature-culture contact zones multiply with each tactile look (Haraway 2008). Walking-with, slowing down, paying attention. What follows is an unedited journal entry written by a fellow participant during the encounter.

We had just wrapped up a pleasant lunch of brown cheese, uncharacteristic sun, and rock seats that seemed custom made for the contours of our back sides. The group split into two to continue our *friluftsliv* activities during our Wild Pedagogies gathering in Finse, Norway. I joined a group that decided to take a meandering path back to the trail. The Norwegian landscape we were enjoying was incredibly navigable. Sightlines extend unobstructed, rocks presented endless opportunities for creative navigation and challenge. Vistas were panoramic.

And the glacier that commanded our attention was atrophying.

Fig. 1.6 Digital Image: Photographer Brant Miller

The group of us was led by a famous Norwegian boat builder who would narrate our walk (not hike mind you, Norwegians don't hike), pointing out interesting natural history features and elaborating on the social fabric of Norway as the occasion dictated. I much enjoyed the group meandering, but I also longed to set out and carve my own path. To explore. To see what I could see.

Fig. 1.7 Digital Image: Photographer Brant Miller

I made an appointment with the group and set out to explore a nearby point of interest. Solo travel provides a contemplative space to engage with a landscape. To tune into the natural world and one's own senses. As I rock hopped, meandered, and generally let the landscape and my curiosity set the course, I came to a fork in the path. A decision point. In front of me I could discern a faint game trail. Possibly from reindeer who use the Finse region and its expansive access to suitable plants for browsing for nutrients during the short summer and fall seasons. As I contemplated the way forward, I looked to my left and noticed a rock cairn on a ridge a short distance away.

I was immediately struck with the juxtaposition of the two options. First, the game trail I was on had become faint and was less clear than the path I was originally drawn to follow. I felt confident that it continued in the direction I was going but because of the reduced explicitness of the clues, I knew that continuing to navigate on that particular path would require greater observation, perception, and intuition. As I looked up to my left and saw the rock cairn in the distance, the second option immediately presented to me a known path. A path that would not require more than looking ahead to recognize the next rock cairn, and the next, and next, until the path become evident, or I arrived at my destination. Two choices that required different things from me. That would provide me with two different experiences. The former would require a different way of looking at the world because the path was originally laid by more-than-human organisms and the forces that moved them for their own reasons.

Fig. 1.8 Digital Image: Photographer Brant Miller

What could I learn from tuning into that knowledge deciphered through seemingly imperceptible clues? The latter would be an exercise in binary existence. Do I see a rock cairn or not? Where is the rock cairn? Walk. What happens when we walk in a line? What happens when we follow a beaten path?
One person in front of the next, step after step, mile after mile.

Journal Entry—Brant Miller

Walking-with the Novella

> "... and so they betook themselves to a little meadow, where the grass grew green and high nor there had the sun power on any side. There, feeling the waftings of a gentle breeze, they all, ... seated themselves in a ring on the green grass." (Boccaccio, n.d., 29)

And so, the storytelling in Bocaccio's *Decameron* begins. The year is 1348 and a group of 10 people (7 women and 3 men), plus seven servants (4 women and 3 men), have retreated to a palace outside of Florence to wait out the bubonic plague. As Boccaccio points out, this pestilence has been wreaking havoc on the world. People have been forced into isolation, families have been separated, fear and rumour are running rampant, and as is usually the case, "the condition of the common people ... was yet more pitiable ..." (Boccaccio, 14) than our economically privileged storytellers. All this is, of course, startling familiar in a worlding context through which the authors assembled this contemporary novella.

Over the following ten days, Boccaccio's ten characters take turns each telling a story a day as a way to pass the time. The tales themselves are at times bawdy or gruesome, poignant, or political, folksy, or historical, wide-ranging, or narrowly focused. The gathering of these stories is a means of reporting the news—from ten perspectives. And thus, the Decameron is likely the earliest example of the novella form. The Decameron is divided into sections, each made up of a single day, with ten stories each day. Coming together to share stories is common in many cultures, and the practice of gathering together in longer forms was, in the west, to set the frame for what has become known as the novella and continues to inform experimentation with creative texts (e.g. gathering of blogs, textual collage, and/or bricolage). For followers of western literature this style of short tales, gathered together (for example, think *The Canterbury Tales* or *The Heptameron*) in a longer manuscript, was to set the frame of what has since become known as the novella. Later iterations of the novella might focus on a singular story which found itself in the nether world of length between a short story and a full-length novel. However, the novella form also includes series of stories, tales, or, as in this novella, commentaries done in different voices but gathered around a singular topic, experience, or point of focus—a kind of reporting of different viewpoints around a central theme. It is upon this history that we see our work resting.

Yet the parallels of the novella in this volume to the Decameron go beyond a simple gathering of stories told over the course of a few days. These tales told here are the results of a gathering, and a retreat, of people from around the world. Thus, participants "betook" themselves to a small mountain hamlet in Norway. What drew them together was the shared concern of a global plague commonly known in some circles as climate change, in others as environmental crisis, in still others as mass extinction. There was however a shared recognition that global challenges need increased attention and that each of us, as humans, and the place itself—as glacier, lake, dwarf willow, and ancient spiders—might have stories to tell. Moreover, there was a feeling that these stories—this news—need to be shared. When gathered, it is our collective hope that this novella might form a small but important educational offering.

And so, as we gathered to spend our days outside. We came to meet the local characters—human and more-than-human—and engage with them in various ways. The tales described explore shared encounters on the land, felt impacts, and joined with us in passing the time. In the Decameron, ten tales are told every day and each voice is heard (other than the servants and the place itself) in its own way. And so, reader, we have modelled this journal come novella on that simple form. Even though we have added bits of discussion and linking threads back to Wild Pedagogies and education, we have tried to do that gently. Sometimes we have presented these participant tales in a format that attempts to bring them into a kind of conversation with our own connecting ideas using columns that are set alternatively to the left and then right-hand sides of the pages. This experiment, particularly in Chapters 2 and 3, reflects a desire to be playful and provocative with journal entries, reflections, and images produced by the wild pedagogues over the course of the colloquium. In Chapters 4 and 5 we emerge from this polyvocal dialogue as we continue to walk in conversation with reader. We also try to leave space for the voices themselves to speak as they would, and for the listeners to gather parts they are hearing into their own whole. Therefore, we invite our tellers to "give beginning to the story-telling… whilst all give ear" (Boccaccio, n.d., 29), as Brant Miller does above and our next two storytellers do below.

Fig. 1.9 Marianne's Rock, Pinhole Camera, Photographers, Finse, Workshop Participants

Everything moves in the wind here—
petals, paths, people, intent.
Wind swirls the silt-point of balance
of thousand-tonne stones
left-alones waiting.
Stability so unsteady.

Chris Beeman

Drawn to edgework,
slippage,
friction,
wind.
What stories shape this place?

Marcus Morse

Walking-with Wild Pedagogies

This novella emerges from a Wild Pedagogies gathering hosted in Finse, Norway. Wild Pedagogies is a relatively new idea that arose first in Canada in the form of a graduate course at Lakehead University in 2012. Since 2014 it has been embraced by an emerging international group of educators through a series of such gatherings, or colloquia. The pervading idea of this series of gatherings has been to combine a mode of travel embedded in a landscape with thought-provoking seminars. And this travelling has been punctuated with times for participants to share ideas. It is through these ongoing and immersive travelling dialogues that the ideas of Wild Pedagogies emerged, evolved, and expanded.

The first gathering was a “Yukon River paddling colloquium” in 2014. This was followed by the Tetrahedron walking dialogue in British

Columbia, a Franklin River rafting colloquium in Tasmania, and a sailing colloquium on the west coast of Scotland. The latter colloquium gave rise to the Crex Crex Collective and the book, *Wild Pedagogies: Touchstones for Re-Negotiating Education and the Environment in the Anthropocene* (Jickling et al., 2018a; see also Special Issues in *Pathways* 2016, 2020, and *The Journal of Outdoor and Environmental Education*, Morse et al. 2018).

During the Scottish colloquium it was decided that the next gathering would take place in Finse, at the top of the railway line between Oslo and Bergen. Meeting there also allowed for the possibility to pay a visit to the cabin *Tvergastein* that once belonged to famous ecophilosopher Arne Næss. Finally, meeting in Norway offered the opportunity to connect Wild Pedagogies with the Nordic concepts of *friluftsliv* and ecophilosophy (Jensen et al., 2022). In addition to collecting insights for this novella, the gathering also generated research published in special issues of two journals (Policy Futures in Education, 2021, and the *Canadian Journal of Environmental Education*, 2022).

Wild Pedagogies arose from a convergence of ideas about wildness, education, and a sense of the social and ecological urgencies that increasingly define our times. The story of our geological age is being written in mass species extinctions, catastrophic events, forced migrations, community upheavals, and increasingly rapid changes in climate. Responding to such issues requires a radical rethinking of human relations in and with the world—with education playing a necessarily critical role in any such renegotiations.

David Orr, like so many others, believes that education must change, for "without exaggeration it will come down to whether students come through their formal schooling as more clever vandals of the Earth and of each other" on the one hand, "or as loving, caring, compassionate, and competent healers, restorers, builders, and midwives to a decent, durable, and beautiful future" on the other (Orr, 2017, p. ix–x). What will it take to nurture caring, compassionate, and competent restorers of the earth? Can the idea of Wild Pedagogies provoke opportunities for reimagined relationships and experiences of enlarged human and more-than-human community? And can wilding ecologies of hearth and home, of self, communities, and cultures become a helpful contributor to this process.

Wild Pedagogies began with two questions. First, why is it so incredibly hard to change the educational direction of our times; are we really fooling ourselves when we think we are making progress? Second, what has gone so badly wrong with human relationships with nature under the influence of western cultures? One answer to these questions is unquestioned assumptions. And one word that links our two questions, and these assumptions, is *control*; we modern humans most often seem to believe that we should exert control—that we should be the masters of just about everything. Many readers will recoil at this assertion, and in a sense, quite rightly: "We are doing good stuff; we outright reject the urge to be in control; there are lots of signs of progress". Yes. However, most of us are marinated daily in the cultural values of western societies and it is nearly impossible to simultaneously keep track of the cultural assumptions that press upon us. Even the most conscientious amongst us can miss deeply embedded assumptions that bend our best intentions back towards the status quo and its controlling dispositions.

In seeking to understand the problem of control, Wild Pedagogies propose to challenge existing assumptions, rethink possibilities, push open doors to educational opportunities, expose limitations of current ways of knowing and being in world, and embrace learning opportunities arising from engaging with the more-than-human world. Wild Pedagogies are also about rethinking our relations in the world. They present a desire to invite the places we visit to become integral parts of our work, and to respond to provocations in spontaneous, and at times unforeseen, ways (Blenkinsop et al., 2020; Jickling et al., 2018b; Morse et al., 2021).

In short, Wild Pedagogies seeks to loosen the leash of control a little—or maybe a lot. And it specifically seeks to renew our art and practice of teaching—to make it less controlled by cultural norms; less dogmatic about expectations; and to make it more disruptive and rebellious (Blenkinsop & Morse, 2017). Wild Pedagogies seek intentional activities that can provide fertile fields for personal and purposeful experience without controlling the more-than-human world and its actors, learners, or educational outcomes. In short, it seeks to make education evermore *wild*.

For some, Wild Pedagogies will provide recognition of things they already do. For others it might inspire a wilding of their pedagogies—providing opportunities for students to attend to the wildness of places and themselves in deeper ways. As such, we suggest that it is possible to offer, through experience, a more inclusive and less hierarchical relationship with place. We are not suggesting that Wild Pedagogies offer a necessarily "new" radical idea; rather, that it gathers interconnected and thoughtful ideas to provoke, or remind, educators to think with and through such concepts. Wild Pedagogies is presented as a heuristic—as an agent of discovery—to provoke practice, rather than a predetermined destination. (See Appendix for an overview of some provocations presented as *Touchstones for practice*).

Wild ecologies, as habitats, are places where the diversity of living organisms is rich and sufficiently robust for the entire community to sustain itself—to coexist and, on occasion, to flourish. It is through walking-with curiosity and reverence through such places, that we create wild pedagogical classrooms. Such places, brimming with pedagogical potential, enable learners to walk and to witness life's myriad forms. These journeys of understanding can nurture broad wildings of ecological possibilities and confront the social and ecological challenges of this precarious era of modern humanity's impact.

Walking-with Wild Pedagogies in Finse

The Finse gathering was planned against the background of earlier colloquia, but with a decidedly Norwegian flavour. The group assembled at a mountain hut for meeting, sleeping, and eating. Rather than undertaking a journey, each day's activity radiated from the hut, evoking a Norwegian tradition of *friluftsliv*—loosely translated as "free air living". Rambling freely in the mountains held a higher priority than completing a multi-day journey.

While it is true that the days were mediated, the planning was responsive to the landscape, and what it could offer on any given day. Travelling in subgroups enabled participants to adapt their route, to accommodate the vagaries of place and pace, to indulge in a deep, slow listening and unveiling of the fragile wild ecologies of the mountain. Thus, there was the presence and possibility for a little wildness in each day's activities.

The Finse gathering allowed opportunities to examine Wild Pedagogies from a variety of perspectives and across a range of discourses, the results of which are published in special issues of journals. However, for this novella a more experimental form of writing was sought. Participants were encouraged to be creative in their writing style, to embrace their emotions, and to tell their stories alongside, and in collaboration with, this place. Though they vary in form, from poetry to prose essays, to images, they are all stories that recount the "news" of their experiences. As in the original novella, the *Decameron*, there is a plurality of voices—or versions of the news. And, again like the original, the stories unfold over a time during a retreat.

In an important footnote to the evolving story of Wild Pedagogies, this novella is striking for its attention to, and co-living for a short time with, Glacier. Participants were drawn to witness and weep for Glacier as they walked within the landscape. As such, we speculate that this represents the emergence of a landscape shaping the research. This gathering also acknowledged the strong historical traditions of ecophilosophy in Norway. Hiking to the revered mountain Hallingskarvet offered a unique opportunity to revisit the work of the philosopher Arne Næss. Here we literally sat in his chair, experienced his domain, and grasped more deeply his emergence as a thinker, a writer, and dweller of place. All this will be revealed in the following pages of the novella.

Theoretical and pedagogical possibilities framed our walking and thinking with Glacier. As we travelled together, we compiled the novella and we sought to explore what it means to be human in the shadow of a melting glacier. We walked a social imaginary, supporting each other to stretch individual and collective imaginations, as we enlivened alternative visions. By drawing on Wild Pedagogies literature we extended our ways of being-in-a-place. Through shared researching and theorising, we considered whether different, possibly better, worlds are possible.

The essence of this novella are the stories. We feel that they can stand on their own and be appreciated as they are. As with good storytelling, this novella leaves much to its readers. People interpret stories differently. Stories give people something to think with, and a measure of a story's worth will be in the amount of work it can do. We think that for those earnestly seeking to be better teachers, natural historians, environmentalists, or more present nomads in a place, there is plenty of work that can be inspired by this novella. However, in this era of exhaustion, characterised by Braidotti's (2019) description of theory fatigue, we acknowledge readers might just wish to read the stories.

For those who would like a little more, we have included links to some of the theoretical thinking that journeyed with us on the mountain. In doing this, we have set for ourselves the task of experimenting with ways that this can assist in border crossing, becoming self differently, embracing cultural change, and building broad coalitions.

In this novella, then, time expands, contracts, and comes untethered from its more recognisable linearity. Knowing and understanding are unmoored from ideas of rightness, truth, or solely human. The stories are representations, at slight remove, of the author's and artist's voices arising against the encounter itself, at the edge of Glacier. These are not the results of deep reflection and sober second thoughts. Rather, in what follows we hope you will encounter in its immediacy, the pain and joy, sadness, and ecstasy, belonging and alienation, and togetherness and isolation of our encounters. A roiling ball of complexity. Much like the erratics churned across ancient landscapes. We seek to let the expressions of the encounter speak for themselves—much as the characters did in the original novella.

Notes

1. These periods are aligned with Global climatic shifts and have happened between 15 and 23 times throughout the past 3 million years.
2. https://www.ncei.noaa.gov/sites/default/files/2021-11/1%20Glacial-Interglacial%20Cycles-Final-OCT%202021.pdf.
3. https://www.miljodirektoratet.no/globalassets/publikasjoner/M741/M741.pdf.

References

Abram, D. (2017) (2nd Edition). *The spell of the sensuous*. Penguin Random.

Alaimo, S. (2016). *Exposed: Environmental politics and pleasures in posthuman times*. Scholarship Online.

Barad, K. (2007). *Meeting the universe halfway*. Duke University Press.

Bennett, J. (2001). *The enchantment of modern life: Attachments, crossings and ethics*. University Press.

Blenkinsop, S. (2018). On nature as a co-researcher. In B. Jickling, S. Blenkinsop, N. Timmerman, & M. Sitka-Sage (Eds.), *Wild pedagogies: Touchstones for re-negotiating education and the environment in the Anthropocene* (pp. 127–128). Cham: Palgrave-MacMillan.

Blenkinsop, S., & Morse, M. (2017). Saying yes to life: The search for the rebel teacher. In R. Jickling & S. Sterling (Eds.), *Post-sustainability and environmental education: Remaking education for the future* (pp. 49–61). Palgrave Macmillan.

Blenkinsop, S., Fettes, M., & Piersol, L. (2022). *Eco-portraiture: The art of research when nature matters*. Peter Lang.

Blenkinsop, S., Jickling, B., Morse, M., & Jensen, A. (2020). Wild pedagogies: Six touchstones for ChildhoodNature theory and practice. In A. Cutter-McKenzie, K. Malone, & B. Hacking (Eds.), *International research handbook on ChildhoodNature*. Springer.

Boccaccio, G. (no date). *The Decameron* (trans. John Payne). New York, NY: The Modern Library.

Braidotti, R. (2019). *Posthuman knowledge*. Polity Press.

Bringhurst, R. (2002). The tree of meaning and the work of ecological linguistics. *Canadian Journal of Environmental Education, 7*(2), 9–22.

Canadian Journal of Environmental Education. (2022). Wild Pedagogies for Change. *25*.

Fettes, M., & Blenkinsop, S. (2023). *Education as Practice of Eco-social-cultural change*. Palgrave-Macmillan.

Hanssen-Bauer, I., Førland, E. J., Haddeland, I., Hisdal, H., Mayer, S., Nesje, A., Nilsen, J. E. Ø., Sandven, S., Sandø, A. B., Sorteberg, A., & Ådlandsvik, B. (2017) *Climate in Norway 2100*. The Norwegian Centre for Climate Services (NCCS)

Haraway, D. (2008). *When species meet*. University of Minnesota Press.

Jackson, M. (2015). Glaciers and climate change: Narratives of ruined futures. *Wiley Interdisciplinary Reviews: Climate Change, 6*(5), 479–492.

Jackson, M. (2019). *The secret lives of glaciers*. Green Writers Press.

Jensen, A., Presthus Heggen, M., Bob Jickling, B., & Blenkensop, S. (2022). Wild Pedagogies for Change. *Canadian Journal of Environmental Education., 25*, 5–12.

Jickling, B., Blenkinsop, S., Timmerman, N., & Sitka Sage, M. (Eds.). (2018a). *Wild pedagogies: Touchstones for re-negotiating education and the environment in the Anthropocene*. Palgrave Macmillan.

Jickling, B., Blenkinsop, S., Morse, M., & Jensen, A. (2018b). Wild Pedagogies: Six initial touchstones for early childhood environmental educators. *Australian Journal of Environmental Education*. Published online first https://doi.org/10.1017/aee.2018.19

Knight, P. G. (1999). *Glaciers*. Stanley Thornes Limited.

Macfarlane, R. (2019). *Underland: A deep time journey*. Penguin Books.

Malone, K., & Moore, S. J. (2019). Sensing ecologically through kin and stones. *International Journal of Early Childhood Environmental Education, 7*(1), 8–25.

Morse, M., Jickling, B., & Quay, J. (2018). Editorial, Rethinking relationships through education: Wild pedagogies in practice, Special Issue: Wild Pedagogies. *Journal of Outdoor and Environmental Education, 21*(3), 241–254.

Morse, M., Jickling, J., Blenkinsop, S., & Morse, P. (2021). Wild Pedagogies. In G. Thomas, J. Dyment, & H. Prince (Eds.), *Outdoor environmental education in higher education: International perspectives.* New York: Springer.

Muir, C, Wehner, K., & Newell, J. (eds) (2020). *Living with the Anthropocene: Love, loss and hope in the face of environmental crisis.* New South Publishers.

Orr, D. (2017). Foreword. In B. Jickling & S. Sterling (Eds.), *Post-sustainability and environmental education: Remaking education for the future* (pp. vii–xi). Palgrave Macmillan.

Pathways: The Ontario Journal of Outdoor Education (2016). Wild Pedagogy. *28*(4).

Pathways: The Ontario Journal of Outdoor Education (2020). Wild Pedagogies Discussed. *32*(3)

Policy Futures in Education (2021). Wilding educational policy: Hope for the future. *19*(3).

Sandlin, J. A., O'Malley, M. P., & Burdick, J. (2011). Mapping the complexity of pedagogy scholarship: 1894–2010. *Review of Educational Research, 81*(3), 338–375.

Springgay, S., & Truman, S. E. (2017). Stone walks: Inhuman animacies and queer archives of feeling. *Discourse Studies in the Cultural Politics of Education, 38*(6), 851–863.

Tsing, A., Swanson, H., Gan, E., & Bubandt, N. (2017). *Arts of living on a damaged planet.* University of Minnesota Press.

CHAPTER 2

Witnessing-with Glacier

Glacier as Witness

Fig. 2.1 "Glacier edges", 35 mm Film image: Photographer Karen Malone

K. Malone et al., *Wilding Ecologies, Walking-with Glacier*,
https://doi.org/10.1007/978-3-031-63182-5_2

A calm before the storm
An oasis
A silver of silence
    A ripple in the pond
A haven of rock and stone
    sheep and wool
    reflection blowing in the wind

—**Joshua David Bennett**

As we circled around the glacier for several days before approaching it directly, it seemed always slightly beyond reach—beyond comprehension. How might we encounter a place that reaches forwards and backwards in time and space? Such an experience is beyond our day-to-day comprehension of human-scale understandings. We are pushed to extend our understandings of time, of difference, of scale, of ourselves, and of relations. There is a dissonance or disjunction that requires us to search for new possibilities, to reconsider relations.

At first glance, from a distance, Glacier seemed silent, stoic, desolate, empty of life or movement. Yet, we know it is in a permanent state of flux. These fluctuations are in slow-motion, persistently extending, advancing, retreating, and losing ground. Humans struggle to witness the adjustments with our eyes. In Winter, between when the day finishes and a new one begins, it will have inched its way forward. Creeping. In the freezing darkness, small crystals reach across surfaces connecting and entangling like a delicate bricolage of icy mycelium branching and knitting fine crystals. And, when the cold passes and the warmth of the sun rays bounce off white surfaces, Glacial ice defies definition as it simultaneously flows and is fixed. The ice sheet melts, pitting its surface and deepening its caverns.

Every moss rock
with a story
to tell. Get Down
here where you can
really see.

Stacy Boe Miller

In approaching Hardangerjøkulen glacier we do not feel the temperature drop as one might expect. There is no uni-directional movement of cold radiation toward those approaching. It does not intentionally hit us; rather, we step into it—we are drawn to it. And we do so with all our senses, seemingly before we expect to. In a moment we silently cross the edgework of Hardangerjøkulen we enter its world. As Macfarlane (2019) suggests, "when you move into a glacier, you enter *its* space. Sound changes, temperature drops, danger grows. The cold comes at you not in the form of fingers, probing, but as a cloud, an aura that surrounds you and settles in your core: *You're in my space now*" (italics in original, p. 355). There is an awareness of the coldness emanating from the glacier, but also of a coldness that appears from behind, enveloping us, seeping into us as we enter its space. The silent coldness enters our bodies and soaks up day-to-day sounds.

"The glaciers change all the time… The colors are always changing… there is always new colour in the glacier. The black glacier, the blue glacier, the white glacier, the grey glacier, all these colours! Better put on a jacket for the blue glacier." (Jackson 2019, p. 76).

Fig. 2.2 "Frozenly friction" 35 mm Film image: Photographer Karen Malone

Robert MacFarlane describes his time on Inylchek moraine: "I could hear the glacier running through its repertoire of sound effects; the slither of blades of slate slipped over one another, dislodged by an adjustment of the glaciers massive machinery; the groan, from deep inside the glacier, of ice separating from ice" (2003, 108).

Coughing

Coughing the mournful cry of my vanishing future.
The raucous coughing of Glacier joins the cacophony of whispering, creaking, dripping, roaring, and trickling water. The roaring of rushing water, heralding a merciless countdown. Standing on the edge, Hardangerjøkulen calls out across the melting ice dome. Rising higher and drawing closer, I anticipated a deathly quiet. A settling. I would crawl up on to her, feel comfort by resting in her motherly belly. But as shared bodies clambered over icy edges the calm and silence were broken by a thundering crash, the rending of a mighty glacier sending forth its limbs. The sounds of her dying throes calling out to the world. The tick, tick, ticking as the hands of a doomsday clock move in time with my hastening heartbeat. She coughs again, vibrations shudder through our bodies and we wait mesmerised. We are held in the slowing, the waiting, the pause, and then the crashing of ice splintering, thundering, deepening her fleshiness. The awkward uncomfortable knowing. She is ruptured, she is in pain. Parts of her are tearing away. She is calving deep, deep in her belly a large slither of ice is now rushing along and joining others in the turbulent underground river flowing beneath our feet. Pieces of her will resurface and rush along, a wild tempestuous body of water we will pass over on our journey home.

Coughing the mournful cry of a dying Glacier.

Fig. 2.3 "Glacier landscapes" Film image Photographer Karen Malone

I have never understood these landscapes. The previously glaciated plateaus and fjords of western Norway, or even the quartzite mountains of south-west Tasmania where I grew up, have never entirely made sense to me. I have lived and walked in both but not appreciated their frictional becoming. Not until I heard this grinding. And it works at me as well; the experience does not glide easily past.

I have understood stone as immutable. And yet here is a phenomenon of water and time that works to grind a landscape into being across crystallised imperfections. A meeting of stone and ice, friction, and mass. Plateaus, tarns, and valleys in formation. How is it that frozen water can animate seemingly impenetrable rock? And how can my species, myself included in coming to experience this place, so easily cause such destruction of this landscape?

Reflection – Marcus Morse

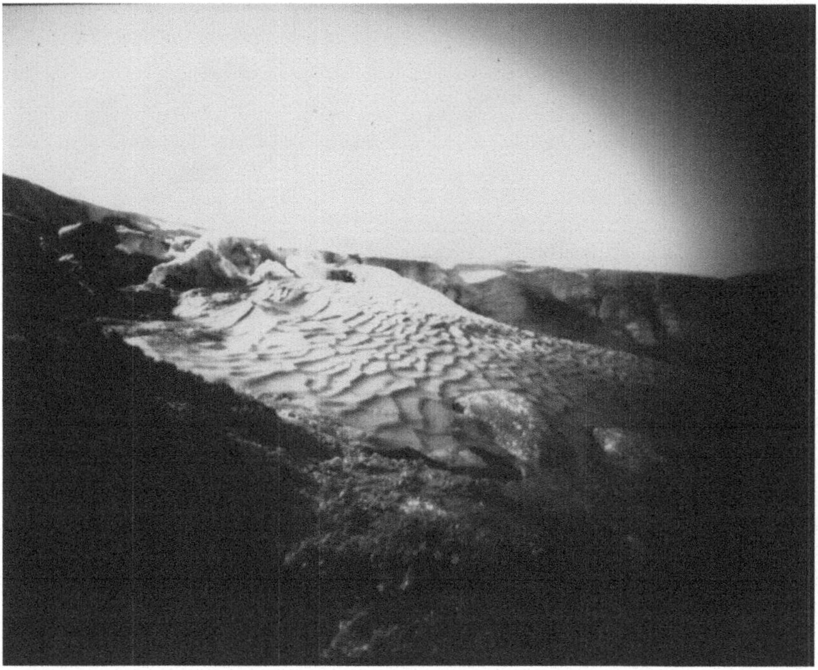

Fig. 2.4 I have seen a ghost, Pinhole Camera, Photographers, Finse, Workshop Participants

Jan Zwicky peers into the geology of Norway and writes:

> Evil is not darkness,
> it is noise. It crowds out possibility,
> which is to say
> it crowds out silence.

(Zwicky, 1999, p. 32)

Fig. 2.5 "Under the Glacier" Digital Image: Photographer Polly Knowlton Cockett

Andrea and I went a little further together and then we were poking our heads under the glacier so… put it that way, I feel that photograph is full of everything we have been talking about. So I don't, this is very scribbled, it's nonlinear, all over the place, so I don't know how this is going to come out but I've been thinking about the primal phases and interphases that we have encountered here, not unrelated to what you are mentioning as—even the sensorial connections with quantum physics, looking at the photograph or out of the window we see, all of the phases of the mineral water and especially when we are up there, we are seeing it in its solid form as ice, we were seeing the vapour, we were feeling the humidity depending

on where we were standing, depending on the temperature and how that affected the air movement, and all of those things.
We were looking at the glacier in some ways as a static moment, in time, but it is rapidly changing. It is representative of the dynamic processes of the landscape, of the weather. The photograph is showing us light and dark, is showing us translucency and transparency. Ah, reflection and refraction and absorption of the light, again, the movement of the water and the ice, the timelessness that we feel. The rocks that are in the foreground are represented as the bedrock of billions of years even through its own story of which we saw a fair bit of on the way up. Ah with folding and crystallisation and the erosion. Also, the erosion happening here where it is all turning back into sediment and glacial silt and all those different phases of solid and liquid that some of us got caught in up to the top of our boots in the mud and (laughs with others) calling for help! (laughs) literally! Being taken by the mountain, it grabbed you by the feet. Being taken, yeah, and then we were, I was talking to Marianne about, she was referring to it as nature, as trickster, as we were coming up to the second glacier, and it was cold because the wind was coming off the ice and it was dark and it was cold and I was feeling cold and there was still some sunlight over there, and it was only going to be two steps to get to the sunlight but my steps took me into the abyss where I could find no purchase to get out.
The sounds, people have been talking about the dripping, the rumbling, the sounds of the wind. Ah the carving off, this whole dynamic thing that is happening up there, the motions people were talking about. The grief, the vulnerability, the love, and the joy um all of this in these processes of conversation and contemplation and communing that we were having in that accordion sign of coming together and moving apart, um, crossing the bridge in the morning with the river coming down and really, I was certainly noticing, someone mentioned it. The different sound of the water when we were crossing the bridge in the afternoon because the temperature had risen and there was more water in the afternoon and that rushing and that pulsing of the day, ahh, the vibrancy that you've been talking about. Anyway, it was pretty powerful for me - all of that.
Audio Transcription – Polly Knowlton Cockett

Scaling

Hardangerjøkulen distorts scale—and it reveals scale beyond usual comprehension. In our day-to-day lives we seem to be able to block out problems of scale as seemingly 'taken-for-granted'—because to acknowledge their reality, to witness them, to be implicated, might force us to act. To stand under the stars is to witness something beyond a measurable scale—yet we can reduce it to our own scale. To conceive of it as 'our night sky'. It is, perhaps, a way of coping—of not needing to act, of not needing to be a responsible part of the world.

In Hardangerjøkulen we hear scale—as mass transforms understanding. The sensorium and self are engaged differently than perhaps the usual experience or conceptualisation of scale. We are immersed in the place. Ice operating under a scale that defies human imagination grinds away at seemingly impenetrable rock. It is not the same as being immersed in a forest—we feel, smell, and taste the realities of scale differently. We hear it in the moment, as solid rock becomes fluid, and we are forced to rejig our comprehension of what is possible. To understand a difference of scale: not the difference of one 'thing' from another—but of one 'possibility' from another.

And in Hardangerjøkulen we see scale—perceived in a ghostly blue. We sense a material denseness, formed over time that returns light to us after its journey through histories, atmospheres, and long-dead creatures. The blue of Hardangerjøkulen, like other glaciers, is the result of snow and ice being compressed over time, of being transformed into glacial ice. Time appears as a blue that draws us into the underside of the glacier. Into the glacier itself.

Ice bergkrystall

For *Hardangerjøkulen - the glacier*
Hardangerjøkulen: a frozen hyper-chord
massively distributed in spacetime
(relative to warm-blooded beings)
resonates within a geologic singing bowl
for millennia—grinding crystal into song.

Hardangerjøkulen: gives zero fuck about
the Anthropocene or some tone-deaf tribe
who thinks only itself well-versed. the glacier
is lost in reverie again, composing life forms
& blue: it cares only for music—it *is* music.

Hardangerjøkulen: like you, knows the end
is coming; knows everything is calving apart;
knows the sun was always going to annihilate
everything—everything you ever loved
will evaporate: just wisps of grey, silence.

Hardangerjøkulen: does not dread darkness
or torrential dissolution of self: the glacier
does not retreat—it merely winced, reflexively
at the sudden blast of white noise, the crowds,
the hot cacophony of programmatic optimism.

Hardangerjøkulen: like you, does not know
what it all means or how to feel about death
by deaf ears—it seems cruel, inhuman to kill
a composer with noise: it seems evil. But it is
late in the day & all the glacier knows is blue.

Michael Dé Danann Datura

It was a cold, dark day, one of many with high winds and rain, so much that vision was impeded. But over my right shoulder, on the less windy side, there was a glacier. And in this cloudy day, it didn't glisten or shimmer or shine like you hear about, but it did glow. It glowed with a colour that I had not seen before, and that became part of me. The colour was a kind of blue, but I knew that though I had seen glaciers before, I had not seen this blue. This was not the characteristic turquoise of ice and water, but a deep grey-blue of old age. A blue layered over with centuries of silt amidst snow, that perhaps I recognized because I have passed a point in life that allows me to see over the crest of increase, to decay. This blue seemed familiar and already known to me, though I knew that it was a blue I had not seen before. I could try to describe it but could only truly say that it was the colour of a certain Norwegian glacier seen at a distance on a cloudy and rainy day.

Yet, I can say, it was faint, the faintest, blue. So much lighter than the soft edge of horizon at dawn, where colours can just be discerned, and where they emerge from darkness. This colour does not 'emerge'; on the contrary, its power comes from being masked by other tones; it is still capable of glowing, yet with other tones like the grey silt overlaying it that somehow permit the internal glowing to occur. I know that you can't photograph this colour of blue that glows through grey from a distance: I have tried. You can photograph a deeper blue up close, but this is not the colour I am thinking of. Nor is this deeper blue more real because it is more proximate. Like few other things now, in the era of the ever-photographed, it is ontologically untranslatable. It cannot be represented with materials that are not itself. The wind and the rain that are between me and the glacier are part of its colour.

Journal Entry – Chris Beaman

Glaciers have an aliveness that western language seemingly searches to describe. For visitors to a place like Hardangerjøkulen such descriptions are inevitably limited by culture, vocabulary, and time in place. Ice itself has an aliveness that is often intimately enmeshed in localised language and life, as Remaud (2022) highlights of Inuit words to describe changing states of ice, confirming "that language constantly registers the modified states of ice" (p. 102). Yet language draws up more than that which is literally described—it can invoke felt resonance: "Glaciers are Chameleons. They take on easily the colours of the sky, the pending rain, people's culture and history, a young man's hunt for economic stability, an artist's dreams." (Jackson 2019, p. 77).

Bluing

Light penetrates my denseness, reflected to those approaching as a ghostly blue, forged over time and under gravity. Snow falls on my upper slopes, watery crystals gently settle atop one another trapping air in layers between. This lightness and pallor reflect sunlight back in its full spectrum as whiteness. But with accumulating seasons, as snows continues to fall on my limits without fully melting, there is a build-up of form. Inevitably the weight of gathering snows act to compress crystals below, as denser layers are formed and evermore weight is amassed over time. Compression continues apace.

Over time, my density increases as the air spaces between crystals are sealed off and layers are transformed into glacial ice. With increasing time and weight, isolated air bubbles eventually disappear, squeezed, and compressed over time into extended crystals of my solid ice formations. Without air bubbles, light is no longer easily reflected outward, and penetrates further inward. As this light travels deeper into my crystalline structure, lengthier red wavelengths are absorbed, leaving remnant blue light spectrums to be perceived by human eyes on the return journey. This ghostly bluing, then, is a view into time itself. Into compression, gravity, and extant histories.

For human scientists interpreting the past my blueness holds memories of times past, volcanic eruptions, the industrial revolution, forest fires, and of warming temperatures—sealed off in compressed air and finally released at my edges. As bubbles are reimagined, air from thousands of years bursts forth, an offering and an undeniable witness—of time, leaves, air, and spiders past. My bluing testifies to possible past-present-futures. To have such a witness is one thing, but to listen, and act on the evidence, is seemingly another.

"[Ice] remembers and it tells-tells us that we live on a fickle planet, capable of swift shifts and rapid reversals. Ice has a memory and the colour of this memory is blue." (Macfarlane 2019, p. 338).

Un/Knowing

Glacial edges cleave into the milk-grey river I feel dumb, she says
Inside ancient ice, memories of endings yet to come Her only
language was light, but she couldn't speak it
Their feet stutter over gravel coded with grey
I don't say everything I know, he says, because I don't know
everything I can say
Wind translates into touch what cold articulates in ice
He deliberates theories of luminosity, while she glows mute beside
him
Inner kilometres of ice tighten with unknown knowing
This is the only way I can speak to you, she says, placing a stone in
his hand in her mind
In white–blue hallways, they stand — heads empty — bodies bright
We need new ways of knowing, he says, listening to patterning drips
A radiance runs through everything
Mountain rocks have always already been alive, she says
Where her hand weaves through water and silt, their knowing
overlaps True things hurt, she says — and heal, he thinks
Sunshine twines cold thread to the edges of frost
Between them, frictions of light

—Estella Kuchta

Messages in Ice

Humbling
Inspiring
Heartbreaking
These sensations pierce my heart as I stand next to a glacier in its full remaining glory, witnessing it melt rapidly right before my eyes.
Humbling – I am small, young, vulnerable in the face of the landscape that has held this glacier in its cradle for so, so long. One wrong step and I am either up to my waist in glacier mud, under a falling chunk of ice, or in a fast-moving glacial river. Presence and awareness are needed.
Inspiring – The power, glory, beauty, and mystery of what this glacier holds is mind-blowing and awesome. Deep gem-like blues weave into the ice like veins. Swirling textures and shapes grace the ice telling a story I can only guess to interpret. Intense rushes of water and splashes from the glacier as it calves cause me to stand still and witness this wonder. Awe and appreciation are needed.
Heartbreaking – This beauty, power, history, and magic that is glaciers everywhere is disappearing as I stand here. Ice melts, dripping water in a rhythmic soundscape mixed only with the occasional splash and the rushing of the river coming out from underneath the massive ice. This beautiful and wise being, who has seen many more years than I can ever hope to, is leaving this earth. It is transforming its being into water, leaving behind its current form. Tears and resolve are needed.
Humbling
Inspiring
Heartbreaking
I stand on this earth as a human wondering how I can hear the wisdom of this glacier and carry its message of presence, awareness, awe, appreciation, tears, and resolve to my species as we attempt to chart a new way forward.

Journal entry —Deb Matlock

Fig. 2.6 Snow 2, Pinhole Camera, Photographers, Finse Workshop Participants

Walking in this landscape, moving carefully across the glacial mud toward the sounds of dripping water and grinding blue ice across stone. I feel in explicably drawn to the fluid edgework of rocks drifting beneath my feet.

Reflection—Marcus Morse

Fig. 2.7 "Losing myself" 35 mm Film Image: Photographer Karen Malone

"Present day Hardangerjøkulen is in a very vulnerable state, and our study of its history over the last several thousand years shows that the ice cap may change drastically in response to relatively minor changes in climate conditions", writes Åkesson.[1]

Ice Castle Retreat

White castle, aglow.
Layered blue, a nod to the
sky; rounded edges.
Your weeping reminds
me of my own, dear glacier.
Restitution offered on
a clear day. The destruction
we carry, akin to tragedy –
have I fallen in love?

A delicate arch beneath your feet
where I dance today
may be gone tomorrow.

I beg this deep bellied goddess
to linger despite our heated home;

The castle of ice has
melted to stone.

—**Erika Kazi**

Fig. 2.8 "Meltingness", 35 mm Film Image: Photographer Karen Malone

Glacial meltwater
rushing the past
past

—**Estella Kuchta**

I've seen a ghost.

—**Erika Kazi**

Meltingness

In constant motion, a river of ice, forming, reforming, and dissolving. I drift under and past the feet of those approaching from below. My *meltingness* is a movement from solid to liquid, as the weight of memories are released across the flatlands below. Compressed under my own weight, I carve out valleys, crushing, grinding, and carrying forward, displacing solid rock as I slide on my own meltwater. Seemingly incomprehensible to those approaching—frozen water tearing and shearing away hardened rock walls.

Ice. But this is only a temporary state, hanging in the balance. Difference occurs and rhythms accelerate—events provoke a precarious movement. High up on my exposed surface water droplets slowly form and slide together, gathering into pools warmer than my icy surface. Meltwater freed by the sun and gravity from its slow-motion movement in its state as solid ice. Shallow indentations melt into ever deeper pools, forming funnels that carry moisture further downwards. Crevices groan ominously, opening up. Near vertical on a sliding incline towards my own lower reaches—steadily growing, spilling down, increasing in flow, and creating well-like shafts or moulins, that penetrate deeper and melt downwards. As I move ever downwards, I accumulate more ancient ice from deep within.

Eventually I escape from crevices and spread into the chaotic friction zone between ice and rock, lubricating a glacial sliding, carving around corners—moving with an inevitable rhythm drawn downwards. Dripping at times at the edges of an exposed underbelly, and roaring past at others. A milky torrent emerges at my edges—a liquid mix of dust, rocks, and memories to be spread across the landscape. Time and memories swirl, lost, seeping, and disappearing under the approaching feet.

The ecophilosopher

"The light of the Enlightenment was not to be blue—not full of misplaced and otherwise abused objects, displaced spectators, short-wave blurred edges of historical lament. It was to be the white light that as we know, but cannot see except refractively, contains the entire spectrum—the light of programmatic optimism: Utopian light. The light of hope, with its dependence on undependable pleasant surprises is, on another hand-to-mouth, blue. It arises out of a strangely luminous pessimism, late in the day. Blue light to be perceived as light at all requires an oblique angle of vision, an averting of the eyes from the glare of our cultural foreground toward the silent periphery that is our future." (From *Blue Notes on the Know Ledge*; Retallack, 2003, p. 77)

Fig. 2.9 An ecophilosopher: Photographer Karen Malone

the ecophilosopher does not enter
the glacier lightly—he has been
contemplating grey for days.

wending his way through
boulder-strewn fields;
listening to the striate lectures
of a thunder-wise massif;

gossiping with smoky quartz
in the back of class
about grey rainbows & friluftsliv.
his dark thoughts, little clusters
of crowberries, pop almost purple
in this alpine light—almost blue
at such an oblique angle.

he tells them:
fieldwork is congenial to philosophy
& reminds them Wittgenstein also made
a number of working visits to Norway
to take blue notes.

what is he supposed to say?
that he is researching the liminal space
between dream logic & landscape?

that his methodology is sleeping in a tent
on a bed of arctic dwarf willow?
that his participants only speak in wisps
(if you shut the fuck up & listen).

no, no—too grey here for them.
besides, not many can keep up his gait,
his strangely luminous pessimism—
the pragmatic epistemology of sandals
in the wilderness.

& now it seems as if every cairn led him
to this place, every grey chunk
of gneiss & granite a sign
pointing
here
the mouth of the glacier.

& now the ecophilosopher—
grey-backed like a hooded crow
grey like a paradox, like lament
grey like mature lovers greyed
from a lifetime of blurred edges
—enters.

the glacier told him:
you are ready now—
ready to be annihilated
by blue.

Michael Dé Danann Datura

From a western perspective we map and understand our world within delineated time, such a taken-for-granted understanding enables us to map spatial wild-lands and "domesticate" them, because we set those things on known time (controlling sea faring time enabled expeditions of "discovery"). We even try to regulate time through geological ages. To enclose time and colonise it; to potentially take it for granted or feel as though we can manage, organise, and influence it. And in so doing, in a moment, we differentiate between one thing and another spatially—as existing in a series of linear moments. We are able to split some "thing" from ourselves because we do not necessarily see things as fluctuating or changing with us in and across events in the same timeframes.

Maybe there is a difference in the experience of not sensing time as a linear series of frames that have differences of degree between frames—that we might even experience the idea of difference differently. When we affectively experience a world of different scale, for example, such as the lived experience of being at Hardangerjøkulen we are implicated in events. Deleuze, in writing on Bergson's understandings of time, highlights implications of thinking temporally rather than necessarily spatially:

> Take a lump of sugar. It has a spatial configuration. But if we approach it from that angle, all we will ever grasp are differences in degree between that sugar and any other thing. But it also has a duration, a rhythm of duration, a way of being in time that is at least partially revealed in the process of its dissolving, and that shows how this sugar differs in kind not only from other things, but first and foremost from itself. This alteration, which is one with the essence or the substance of a thing, is that we grasp when we conceive of it in terms of Duration. In this respect, Bergson's famous formulation, "I must wait until the sugar dissolves" has a still broader meaning than is given to it by its context. It signifies that my own duration, such as I live it in the impatience of waiting, for example, serves to reveal other durations that beat to other rhythms that differ in kind from mine. Duration is always the location and the environment of differences in kind; it is even their totality and multiplicity. (Deleuze, 1991, pp. 31–32)

What happens when such temporal ruptures occur? When we sense that the time we had contained and measured was only one possibility, a seeming illusion, that we have not surrounded and enclosed measurable time. But rather, time is swirling around us, disappearing in unknown and uncontrollable ways, wild time, is moving in scales beyond our reckoning or control. Glacier time. Might possibilities and imaginings open alternative ways of being in the world? Bergson's understanding of time seemingly implicates us more directly within the world as finite beings. Might the lived experience of walking with Hardangerjøkulen, for example, provide such a piercing—a rupture that potentially calls into question forms of differentiation? And might such possibilities hint at the finiteness of relations, and the appalling ecocide of climate change? We suggest the answer might be yes, and that such a view is supported by the walking-with, writing-with, and thinking-with Glacier.

Richardson (2020), similarly asks, "might opening up affective witnessing to nonhuman entities as productive meaning-making agencies enable new ways of grasping the temporal scales of catastrophic climate change?" (352). Witnessing scale is not just about being removed—but potentially about being more intimately implicated. It can involve feelings of responsibility and a need to act. In witnessing an incomprehensible scale of time in which we are directly implicated there resides a level of responsibility that seemingly requires personal and collective action, but also a comprehension of time beyond the momentary near-scaled human experience.

When we hear from a witness we make judgments—about the story and the storyteller. Listening involves trust—I choose to listen to you, I believe you, I take your words as evidence? We make judgments based on trust in our day-to-day lives—listening to the news, engaging with social media, and hearing stories from friends and family. And what if we were to extend the idea of witnessing beyond the human? What does how we listen say about trust? How might places appear as witnesses of the past, present, and the future? Again, conscious, or not, we make judgments based on social imaginaries of trust in places.

If witnessing provides a first-hand account—then Hardangerjøkulen presents an undeniable account of bringing the past into the present, then seeing the future. As bubbles of gas emerge from ice, sealed off air from thousands of years ago and crackle into the present. The glacier preserves a witnessing—of leaves, histories, warming climates, changing atmospheres, and spiders past. And, as Macfarlane (2019) writes, "ice has a memory. It remembers in detail and it remembers for a million years or more.... It remembers and it tells—tells us that we live on a fickle planet, capable of swift shifts and rapid reversals" (pp. 337–338).

Åkesson says "The topography and present climate is such that we soon expect yearly net melt over the entire ice cap. This has already happened a few times in recent years. In the near future we expect this to occur much more often, and with this, the demise of Hardangerjøkulen will accelerate".[2]

As the ice of Hardangerjøkulen melts it reveals past histories. Are we worthy of its trust?

Climate change is happening on an incomprehensible scale, leading Richardson (2020) to ask "how, then, might the scale of climate change be meaningfully grasped? And, to sharpen the question, how might it be witnessed?" (p. 340). An effective response cannot rest entirely on cognitive understanding—it must also embrace sensorial, affective, and embodied understandings that take into account alternative scales—and with it—alternative relations. We share this witnessing by walking with/ in places—this place. By being there and feeling we are directly implicated—it makes a difference: "the affective witnessing of scale is entangled with cognitive comprehension but is more than simply its spark: affective witnessing describes a modality of relation of scale of global catastrophe that runs alongside but is not subordinate to cognitive understanding" (p. 341). In witnessing climate change we are called to move beyond our usual ability to comprehend scale. We cannot respond at the same old scales; and that presents a problem and an opportunity.

To witness is about more than observing, being present, it involves an injunction to act, even if action seems impossible; that is, to witness an event is to take some responsibility—to be implicated in the event itself (Peters, 2001). Tracing the complex history of the intricately tangled practice of witnessing, Peters shows the necessary unreliability of bearing witness, the inherently contestable nature of the speaking of testimony. Witnessing, incarnated in expressive and actionable form as testimony, reaches into the past in order to consider how to make sense and shape or speculate at least about potential futures. "To witness is to wish that the record of the past was more whole and to grasp this lesson now is to live vigilantly, to make the present worthy as we imagine contemplating it from a future point" (Peters, 2001, p. 722).

To be engaged in *witnessing* means understanding witnessing as fundamentally, even constitutively, affective. All witnessing in this sense is relational and embodied. Affective witnessing of an anthropogenic process such as this melting Glacier is to be entangled with a knowing that exists beyond our easy comprehension: affective witnessing describes a modality of humble being-with, intermingled with an awakening.

In years gone past Glacier would grow each winter more than it melted each summer; moving on the ground under its own weight; and it would have been partially covered with a layer of snow.

"Today the ice is more than 300 m thick at places, which may sound like a lot. But the implication of our study is that if climate warming continues, this ice cap may disappear before the end of this century. I don't think most people realise how fast glaciers can change, maybe not even us as scientists. if Hardangerjøkulen melts away completely, it would not be able to grow back again given today's climate", says Åkesson.[3]

Now as climate change disrupts this intimate balance between glacial growing and its shedding, the seasonal rhythmic release of ice that spills into the lands and river.

Together we stood on Glacier hearing their stories and witnessing their stoic and fatal acts of transformation.

Glacier is melting beneath us.

Notes

1. University of Bergen. (2017, February 14). Norwegian ice cap 'exceptionally sensitive' to climate change. ScienceDaily. Retrieved November 28, 2023 from www.sciencedaily.com/releases/2017/02/170214094034.htm.
2. University of Bergen. (2017, February 14). Norwegian ice cap 'exceptionally sensitive' to climate change. ScienceDaily. Retrieved November 28, 2023 from www.sciencedaily.com/releases/2017/02/170214094034.htm.
3. University of Bergen. (2017, February 14). Norwegian ice cap 'exceptionally sensitive' to climate change. ScienceDaily. Retrieved November 28, 2023 from www.sciencedaily.com/releases/2017/02/170214094034.htm.

References

Deleuze, G. (1991). *Bergsonism* (H. Tomlinson & B. Habberjam, Trans.). Zone Books.

Jackson, M. (2019). *The secret lives of glaciers*. Green Writers Press.

Macfarlane, R. (2003). *Mountains of the mind: A history of fascination*. Granta Books.

Macfarlane, R. (2019). *Underland: A deep time journey*. Penguin Books.

Peters, J. D. (2001). Witnessing. *Media, Culture & Society., 23*(6), 707–723.

Remaud, O. (2022). *Thinking like an iceberg*. Polity Press.

Retallack, J. (2003). *The poethical wager*. University of California Press.
Richardson, M. (2020). Witnessing the anthropocene: Affect and the problem of scale. *Parallax, 26*(3), 339–359. https://doi.org/10.1080/13534645.2021.1883298
Zwicky, J. (1999). The geology of Norway: Poem with introduction. *Harvard Review of Philosophy, VII*, 29–34.

CHAPTER 3

Weeping-with Glacier

Our method for this chapter involves taking things at a walking pace. The slow, individual, movements of the human as they, with no confirmed aim, wander and wonder flaneur-like through a gallery *en pleine aire*. It is this outdooredness and expanded sensory aspect that hopefully plumbs the affective such that our visitor-friend is not just an observer but is re/immersed in an encounter that combines the words, the place, and one's own history and experiences. So, just like its novella progenitor, the Decameron, welcome to the storied responses of a group of people, come together in a place in response to a plague. Stories shared through a variety of mediums and emblematic of diverse responses they revolve around our second day. A day where a particular glacier became a central participant, demanding our attention as human visitors.

Amongst these installations and exhalations we, as curators, have seeded the space with provocations—small paragraphs of text on a grey background. Little moments of fecundity that have made themselves unignorable as we worked with these pieces. We hope they too will speak as you, our invited guest, encounter them in your own tuned in, tuned out, embodied, recognising uniqueness. We have entitled this day, this collection—"weeping"—for grief and the redistribution of water across our faces and the face of the planet appeared as a central theme but don't let that impede your own emotional engagement. For to weep is never

K. Malone et al., *Wilding Ecologies, Walking-with Glacier*,
https://doi.org/10.1007/978-3-031-63182-5_3

simply about grief; to embrace and encounter another's words is never simply about grasping a singular understanding.

Walk, wander, wonder, weep, whatever ... but at the very least, listen.

Fig. 3.1 "Midtdalsbreen" 35 mm Film Image: Photographer Karen Malone

Midtdalsbreen Introduction—Marianne's Story

> This is a small part of my story from when I introduced the Wild Pedagogies group to the glacier Hardangerjøkulen in August 2019. Midtdalsbreen, or Mid-glacier, is just one arm of the Hardangerjøkulen glacier that is a looming presence above the small mountain train stop of Finse, on the Oslo-Bergen line. It was also the most accessible arm to reach on foot. When we arrived at the glacier, I introduced our group to my history and presence in this place. The story presented below is based on an audio-file of my presentation at that time and was recorded by one of the participants. During

my presentation, participants were invited to share their own impressions. Many did. Before approaching Midtdalsbreen, we talked about wild pedagogies—in theory and practice—and nature, we shared a meal, and then in the afternoon we walked up to its edge. I believe that this gradual and slow progression helped us tune in to our target. However, there were many distractions along the way and the tuning in seemed to be more branchlike, with thoughts dwindling off like limbs on an old, twisted oak, rather than like a stream that gathers smaller dribbles of thoughts that continually build volume along the way.

Although we were heading towards the same place, maybe our thoughts were not?
Shortly after reaching Midtdalsbreen, we gathered on the fresh moraine, and I started to welcome everyone to this, special place:
I would like to welcome you to the land of the ice, the spiders, the mosses, the lichens and the glacier.

And then, I found myself telling the story of a reconnaissance visit two days earlier:

We approached from the other side, trying to find some place where we could access the glacier. And, what met me wasn't really the beauty or the magnificence or anything like that. It was just an extreme sense of sadness. Because this glacier is so much smaller than it used to be. It used to be large, huge, majestic, magic, it used to go all the way: you can see all the grey, it used to go all the way ... It's a dome glacier, so it lies at the top of the mountains, and it runs... it falls down it's sides, in all these different arms.

And you can't see the dome anymore. The dome used to be towering up, over the glacier. And it's not like that anymore...

Experiencing this, I started weeping.

As a researcher, a scientist, a natural scientist, I could calculate, I could see that OK, At the edge of the mosses at the edge of the lichens, OK, that's where it used to be. It hasn't really withdrawn that many meters, it has withdrawn much more in height. You can see, how high up at the side, the green vegetation starts. So, I can't really, as a scientist, understand, I can read all this, I can look at the maps, but I can't really understand it, if I'm not here. So that's partly why I wanted to bring you here. I don't go here anymore.

Marianne Presthus Heggen

Because of the sadness?

Yes. It's depressing. This used to be marvellous…

Marianne Presthus Heggen

Where Can We Keep Our Sorrow?

Flaking shale gray into our sleep,
our palms, our open
mouths? Do we hide

it deep in mountain lakes
or fly it
out windows like a spring flag?

Oh trespassing
lavender, oh golden
moth, oh surprising
long legs of our daughters,

give us what we need
to look past this grief. Give
us streams where naked
children swim. Give us verses
  to inscribe on our forearms

Fig. 3.2 Digital Image: Photographer Brant Miller

Give us growing
glaciers and rocks
resting on other rocks,
and give us wildness

to tunnel
our bodies to
a height where we can
see it all.

Stacy Boe Miller

Paradox

Intriguingly, the visitor can see these moments of paradox written across the exhibit. Moments where the artists feel themselves as both significant and insignificant, victim and violator, Elder and child, joyous and melancholic, human and more-than. And we notice this desire to resolve these things, to name the singular emotion, to position and summarize the complex others we encounter, to identify and come to a comfortable resolution with regard to the myriad others in, on, and beyond us. Many of whom, despite our best attempts, continue to escape our desire to name, to hold, to solve, to know.

Maybe it is in paradox, and feeling our way into sitting with it that we might escape the tyranny of teleology, the power of progress. For paradox opens the possibility for the other to make demands on how and if they are known. It also makes the human potentially both knower/unknower and object/subject of knowledge simultaneously. Opening space for the world, for dying glacier, for ancient spider to know, to have their beingnesses, becomingnesses, epistemologies recognized. And to become teachers and knowledge carriers in their own right.

The day started out in a circle. A discussion of glaciers, dissolving ice landscapes, sadness. Our energy matched the sky, which was heavy with a grey that seemed to sit atop the world like a blanket. The factual conversation of glacial melt continued as I gazed onward at the sleeping giant in the distance. I pondered time, entropy, and the irrevocable madness of climate-change in the Anthropocene. The glacier's behemoth size was puzzling. Were those arms really part of the same glacier? Could they really droop around the mountain in such a lazy fashion? What a quiet beast, accepting its fate.

Someone mentioned the sound the glacier makes as it melts: a groan of the injured, a roar of the defeated. As I stared over Finse, I felt the anguish of that ageless being across the lake. I tried to imagine its disappearance. Tried to visualize the magnitude of melting. And with the weight of this knowledge, we wandered south towards the retreating ice.

I walked alone, hearing my own footsteps against the rocky ground, a gentle lapping of the wind soothed lake, and a distant murmur of conversation carried on the breeze. Just like yesterday, and the days before, I walked past a colony of swamp wool against the shore. Only today I noticed the numbers had grown significantly and this particular indentation in the land seemed to be full of white tufts of fur.

How awe inspiring it was to witness the land and its various personalities transform around me. Look at the swamp wool as it grows! Watch as it appears day after day, multiplying itself seamlessly. Small tuffs of fur poking out of a harsh, rocky landscape providing a white warmth, like stars in a green sky, bouncing in the moving air. With this, I was brought to tears. In a brief, seemingly insignificant, moment I witnessed, understood, and accepted *change*.

Journal Entry, Part 1—Erika Kazi

Although we were heading towards the same place, maybe our thoughts were not?

I've never been to a glacier before, so I am just desperate to go to it.

I feel really playful. Because, though all these changes happening, though there's all this dark matter, all these devastating feelings going on inside me, inside all of us, in the world, I feel really grateful to enjoy myself around the glacier, on top of a mountain. With, you know, like-minded people. So, it's like taking that dark versus light and find a way to bring them together.

And when I came up here, I wasn't that sad as you are. Because I still feel the power from the glacier, the sounds, and I am happy to be here. So, it's a powerful area. So, so, it's more like, I'm happy to be here, more than sad of what has been.

Companions

Fig. 3.3 Cotton Grass, Pinhole camera, Photographers, Finse Workshop Participants

Cotton friends, wave above tundra
Stems pierce the same moss
I rest upon
sharing the damp—the warm breeze
on our cheeks
Bob Jickling

We bend ourselves
to the sun when we see
the sun. We eat
what the morning gives
we point,
for you,
to where the wind is going
Stacy Boe Miller

Grandmother Glacier

On the journey to the glacier, colleague conversation meanders from fathers and songs to Saami and reindeer. I am pulled towards the current, that mask of human murmurings.

The river is magnetic, drawing me from one tributary to the next like a spider weaving her web. The water finds her own path, slipping with gravity to future lakes.

This Norwegian wildscape is host to the Wild Pedagogies Colloquium, for academics craving query and conversation with the more-than-human. We will linger in this place until the place starts to linger in us.

I arrived to Finse with a bag full of contradictions. A father 7,000 kilometers from his family. Supporting sustainability while flying over the Atlantic Ocean where Greta Thunberg is cruising on a fossil-free sailboat. My grandma's death three months prior. Now, walking alongside this mountain artery, this ceaseless maelstrom of riffles and eddies, pushes at my dam of emotion. What grief am I holding?

During the colloquium's introductions, the nearly thirty participants from five continents had an opportunity to speak to their expertise and current ideas. These academic monologues did not facilitate sharing emotional baggage, nor was this the intent. Thus, my jumbled feelings—about home, family, and environmental catastrophes—remained dormant.

I wanted to speak to the hourglass t-shirt I saw on a fellow airplane passenger. Mountains, trees and rivers wound across his chest, while on the bottom half, toward which the sands of time fell, there sprung angular roads and skyscrapers.

I wanted to speak to my horror at the potentiality of this ominous prophecy walking past me. But Finse had something else in store for me, something I would find at the glacier.

Journal Entry Part 1—Lee Beavington

Fig. 3.4 Grandmother Glacier: Photographer Lee Beavington

The Emotional Paradox of Ice

It is the last day of August, and the sun has made a meagre mid-day attempt to push its way through the dark, damp layers of cloud that have brought wind and rain into our lives for the last 16 hours. We left Finse in the early afternoon after a day full of hugs, good-byes, gratitudes, and plans for next steps. We were tired from 5 days of talking, thinking, walking and, I sense, deep emoting. Feeling into the world and its myriad more-than-human denizens can be exhausting work, particularly if done with the awareness that many are threatened and that places such as these exist, tenuously, on borrowed time.

Journal Entry Part 1—Sean Blenkinsop

Although we were heading towards the same place, maybe our thoughts were not?

So, not to continue the sadness, but if we came back in a hundred years, would we find an esker?

Oh! You wouldn't have to wait that long! Come back in ten years…

This is very ephemeral, in time's way. In ten years, this will have turned into these longitudinal snakelike features.

These will never get to be snakelike like the continental glaciers.

Marianne Presthus Heggen, and companions

Fig. 3.5 Midtdalsbreen Melts/Marianne's Tears: Pinhole camera, Photographer Bob Jickling

This photograph was made using a camera with a pinhole to allow light an entry. There is no staring through a lens, no viewfinder or light meter. There is no checking, no certainty, just attentiveness.

> As this once-domed glacier shrinks, it perishes. Eyes mist; cheeks moisten; shoulders heave. This great icy bit of Earth speaks again. and makes her presence felt. Marianne weeps.
>
> **—Bob Jickling**

Exposure times are measured in half minutes. Instants give way to passages of time. In the interval between uncovering the pinhole and then covering it again with a finger-for-shutter, the glacier thaws and a river of melt water passes in blurry impermanence. Tears are shed. Rock is ground, and then blown by the wind. Ancient life is reborn as ice dies.

> Earth has always been an active presence, but in a role that has often been overlooked and denied in modern thinking. We mostly choose not to listen. However, on this day our little band of pedagogues stood side by side with Midtdalsbreen —in shared fragility.
>
> **—Bob Jickling**

The previous photograph is a co-creation made in the company of blown loess—or rock particles smaller than sand yet coarser than clay. These dusty teachers have embedded themselves in the photo emulsion—having slipped through the back door of our simple camera. Now, their vibrant spoor are reproduced in the centre of the photo, and scattered towards its edges. These traces tell us that Midtdalsbreen is authoring a paragraph in Earth's new geostory—provisionally called the Anthropocene. The black dots remind willing learners that not all meaning arises in human terms.

> And who knew that Midtdalsbreen was a trickster? As she walks slowly up the mountainside, she is followed by a clutter of spiders, grazing at the melting edge. Clever scientists have carbon dated these arachnids at a thousand years, confounding—briefly fooling—humans with their own technology. Meanwhile, recoiling from the

warming valley, Midtdalsbreen regurgitates vegetation that was first deposited on her surface millennia ago—providing ancient fodder for her spider-partners.

—Bob Jickling

Time

Potawatomi scholar and scientist Robin Wall Kimmerer (2013) suggests that for her Indigenous community time is conceptualised very differently from the way the modern west does. She posits that in the west time is like a river. It flows in a linear direction with the past in one direction and the future in the other. There is an inexorable movement to it, a flow. Which also makes it easily positioned as limited, a scarce "resource" quickly passing us by.

However, Kimmerer suggests that a better metaphor for understanding time is that of a lake. Here past, present, future are all mixed and always close by. This tends to tie the community to ancestors and future generations in different ways for they are ever close at hand, while also changing the relationship to time quite significantly. For it never really has an end in the way the river flows away and all are always and already present no matter the when.

For us, standing by the glacier, there was a sense of the impossibility of conceptualising time on the scale of this ice or on the fleetingness of its liquefying. It was also difficult to understand deep time in its geological nature or how these spiders could be 1000 years old. The very embodiment of the past in bodies that are consuming the remains of the ancients. A world once frozen deep, melting out of the ice and becoming life … again.

Oh, to fully accept the inevitable chaos that comes from constant alterations to the world we know. To recognize the role that we play as humans in the extinction of our beloved. To hold the world in our grasp one moment, only to realize that we never really held it at all. To accept the fleeting nature of each moment, each breath, each passing instance of time and beauty.

I stared at the glacier, aware of its ancient timeline. Recognizing its enormous nature and the vandalism it was experiencing, my face continued to wet in emotions of loss, entropy, lack of control. Witnessing the glacier, a forgotten goddess, who was disappearing before me, unresponsive to humankind's selfish hazing of the planet and its processes.

In this devastated state, I tiptoed from rock to rock suddenly hyper-aware of the moss, grass, and mud beneath my feet. Suddenly the world felt alive, and vulnerable. I felt open to a space that was breathing for me, at one with a bitter taste of destruction. My own trauma was laid out here, into the ears of others, into the landscape that shared my sorrow.

The berries, the boulders, and a few kind hearts bore witness to my bleeding soul who stumbled to find some outlet for the most intense form of vulnerability one can share with the world—that of mortality and pain. I spoke of a deep understanding of the glacier's experience—being cast aside for someone else's desires and consumption. Of being raped by someone I had cared deeply for. I fumbled through the grief of arrogance and selfishness. I expressed the struggles of pressing charges to hold my assailant accountable. I was at a loss for answers, especially as I witnessed one of the most desolate sights one can behold: that of a disappearing glacier due to human-induced climate change. In that moment I felt the helpless weight of betrayal.

With a few embraces, the blue-sky tiptoeing into our periphery, and a dancing stride among the hills and mountains, I regained myself. The clouds passed overhead, releasing the weight of the clouds to the open and expansive blue. I was prepared at last to meet the glacial goddess and bow to the carnage that has implicated our lives.

When we finally made it to the glacier, I suddenly felt playful, gracious, and closer than ever to myself and those beings (human and otherwise) around me. The glacier herself did not feel like the broken beast I had envisioned. Instead, I stood beneath a being drenched in the glory of existence as it is and as it were.

I scurried beneath the dripping ice to see the glowing blue colors, to feel the cold kisses of her melting body. The sensation was epic and familiar, like the feeling of being buoyant in cold water; I felt as though I was floating in discomfort. I found pieces of ice; translucent remnants of bubbles, water suspended in time. While the great, frozen being hung itself around cliffs and rocks above me. I felt her weight inching closer. I listened to the roar of the proglacial river, blasting its way into the valley beneath us. Watched as large chunks of ice hung suspended, amid their dive into the moving waters.

Journal Entry, Part 2—Erika Katzi

I have diverged from the group. From my creek bed path, I spy our destination: a white shawl of frozen time known as Midtdalsbreen glacier. The river splits into rivulets, sharp banks hemmed by moss, a verdant flash of life in this sea of grey. The moss offers my feet respite from the uneven Finse stone, yet these leafy depths deceive. Occasional steps sink me deep, drawing me right into the mountain. I am called to slow down. To feel both the mountain and what the mountain feels. These stones are full of stories.

The penultimate ascent is strewn with boulders large enough to reveal epochs of sedimentation. There is a history here that stretches far beyond my settler European ancestry. A geologist's delight. One from our group, an over-zealous rock climber, tries to scale these layers of time. Then we continue to the main course, the place the river is born.

Fig. 3.6 Story Stones: Photographer Lee Beavington

Cameras emerge, insufficient lenses to capture the field of ice staring us down. The glacier is a living fortress of solid rain. A 4000-year-old memory. Cloud-sister. Frozen time capsule. Mountain cloak.

Everyone scatters to find their own meaning. We are called back for a circle, and return with reluctance. While stories are shared about this glacier's retreat, the sound of its disappearance streams into our ears. There is no single drip. No whispers of decay. A storm of tears weeps the glacier away. Huge calves of ice crash with alarming regularity into the unfrozen river. Everyone seems eager to investigate this wondrous tragedy.

Whose tears does this glacier shed? Perhaps the dead lemmings locked in the ice. Or the rock lain bare by the life waters evaporating before our eyes. Or humanity recognizing its collective undoing.

Journal Entry, Part 2—Lee Beavington

Fig. 3.7 "Pieces of me" 35 mm Film Image: Photographer Karen Malone

Although we were heading towards the same place, maybe our thoughts were not?

> *Let's see what kind of birds can I see here? ... there was something else that puzzled me. It was the white fungi. In Norwegian, we call it smoke fungi .. røyksopp ... we found it... they are so beautiful! White, I haven't seen it before up here. Suddenly, from the cabin and up, they are staring at me! Oh my good! There you are! An old friend... Of course ... I have opened these fungi many times before, but I know exactly how they are inside, but, up here I won't do that, because, in respect for them, ...*
>
> **Marianne Presthus Heggen**

We set up in pitched rain last night, scurrying to minimize wetness to person and gear and surrounded by fog that makes visibility limited while giving a feeling of being quite alone. Above us, though not currently visible, hunkering down in this miasma of mist and wind is the Hardangerjøkulen glacier. This grey, white, and deepest blue icefield rests, like a cracked Robin's egg, in a bowl of stone, carved to its own specifications over the last 4000 years. Short tongues of barely liquid albumen leak slowly out the sides as snowy winters have added content to its heart and forced it to push out some satisfactory extensions. From where we sit now, I can see three such tongues of ice reaching towards me.

I can also see raging waters emerging, melt waters drooling out from beneath the ice, seething and crashing sounds that reached us in our tents last night. In the distance, looming behind each glacial tongue is the doom of its imperial summit. Rolling away from us in the distance, towering high above the carved stone cliffs and the massive moraines left behind as ice prepared its nest. From this vantage point I am touched by the glacier's grandeur and seeming timelessness, its slow patience, incredible power, and deep wisdom.

But that is not the only story here. In fact, it is the story of the cursory glance, the cultural cliché, the first-time visitor and the Instagram click.

Journal Entry, Part 2—Sean Blenkinsop

Fig. 3.8 Looking In: Pinhole camera, Photographers, Finse, Workshop Participants

Here are my prayers
Muddy feet of a goddess
Sorry I'm so late
Stacy Boe Miller

Crying ice
Creating emptiness
The End?
Sean Blenkinsop

Colonization

Leanne Betasamesoke Simpson (2017) places heteropatriarchy right at the centre of colonization. Attacking, denouncing, marginalizing, violating, and silencing the bodies/voices of women, children, and the wondrous range of Indigenous queerness was an early and ongoing move by colonizers. Furthering the project of disappearance of culture and land dispossession.

And we ponder, if in colonizing cultures it is not just these gendered binaries but that of human/"non"-human that needs to be named, born witness to, accounted for, undone, and reconceived as well. For Simpson, as an Indigenous scholar and activist, the cultural relation to land is fundamental and assumed. For European colonizers exactly the opposite is true and the attack on bodies and voices of those deemed to be other includes not just women, children, and the queer but also these bodies of Ice, Spider, and Moss we are encountering.

Glacier is Alive

Fig. 3.9 Field Notes—Karen Malone

Fig. 3.10 "Retreat" 35 mm Film Image: Photographer Karen Malone

Glacial till
emotion in the space left behind.
in the space left behind emotion
the space left behind emotion in.
space left behind emotion in the.
left behind emotion in the space.
behind emotion in the space left.
is…
—Maja Zimmermann

This is the connection we share with the natural world, whether we are aware of it or not. The surrounding land has subtle influence on us—through the changing of seasons, the growth of plants, or the lasting changes that creep into our lives. It is this aspect of our ecology that understands us best—the growing of swamp wool and melting of glaciers that recognizes the moments that influence us and the years that shape us.

It is not without deep courage and vulnerability that one can identify oneself with the sorrow that exists in the land and within our lives. It takes opening, perhaps like the glacier's rivets, collapsing caves, and streaming tears, to understand the deep and robust rawness of life: among our greatest sorrows are treasure-troves of joy.

By relating in such an intimate, metaphoric manner with this landscape, I learned about the balance of dark and light in life. If frozen depths of winter did not reside at 1500 meters above sea level, perhaps the glorious blue color of a glacier's underbelly would not exist. Perhaps the mounts of sand and silt would not be there for us to crawl on. Perhaps my own happiness could not be experienced without pain.

And still, as I walked away from Hardangerjøkulen, I found myself balancing on a new precipice of change, acceptance, and adoration. A unique feeling overcame me; I was impossibly numb for someone who was so deeply alive and bursting with appreciation.

I move forward into my life refusing to think myself to death of the burdens of trauma. Choosing instead to observe, patiently, the changes that develop in my life—within each moment, within each decade. Accepting the story that is happening all around me and within me of growth, destruction, and rebirth. Discovering the circular pattern of becoming earthbound; forgiving the chaos, opting instead for gratitude and joy.

Diary Entry, Part 3—Erika Katzi

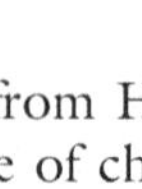

I hold a single drop in my hand. Cold, clear, quiet. A fragile offering.

It strikes me: this tear is ancient, formed before Buddha attained enlightenment. Before the bible was written. Before there were pages to write on. Before humans invented quills, cars, computers and Twitter, this drop of ice formed.

The tear in my hand is not an annual event, brought forth each summer by the meridian sun. No, this moment is 4000 years old. The spiders in this glacial hunting ground that drink these tears, and prey on mummified insects, consume thousand-year-old memories in every bite.

Here, in this timeless place, in my forever memory, my grandma arrives. My grandma carried the family. She showed me the wonder of rivers and rain. She endured when others did not. She held the mountain together.

Yet she is gone. These ancient beings, that by all rights should never leave, are gone, fading into fluvial memories.

As grandmother glacier slips into the past, I feel a well of despair too deep to be filled with mere words of comfort. How to hold this sadness? How to massage this grief into action, so we can flip that hourglass around? Or at least tilt it, just enough, so that for each grain of sand that falls toward so-called progress and development, at least one grain also falls toward ecosystem integrity and social justice.

What is the solution? I'm not sure. But it won't be found in roads and cloud-cleaving skyscrapers. No, we will hear the answers from the murmuring rivers and trees. In the heart of the mountain. In the whispers of grandmothers and glaciers before they pass out of this world.

Journal Entry, Part 3—Lee Beavington

Fig. 3.11 A fragile offering: Digital Images: Photographer Brant Miller

Opacity

We are struck by the translucent blue of the glacier's heart. Ancient water squeezed into relation with fine sand over time producing a colour that exists nowhere else. What might it be called? What does it mean that the word blue seems so inadequate to cover the incredible range of shades the colour possesses? What don't we see because we can't express it or because it has been deemed unimportant even though it is right in front of our eyes.

Cree scholar Jarrett Martteneau (2014) has called this opacity. The inability, intentional unwillingness to see what is directly in front of us. Often applied by the colonizer to not know the suffering of the other. Or by one culture to dehumanize another for to not encounter, to avoid dialogue, to remain separate from is to further the cause of ignorance, to continue to make opaque. Here the ice and sand are pressed into relation and yet light passes through an ancient lens that is way less opaque than the younger parts of this glacier.

Although we were heading towards the same place, maybe our thoughts were not?

> *I don't know. Maybe that's how you presented it … I feel emotional and I understand why, I understand why climate change is upsetting… And I see how important it is to contextualize…*
>
> *I wouldn't mind, to have to sit and spend a little time by myself for a little while….*
>
> **Marianne Presthus Heggen, and companions**

Three days ago, our encounter with this living being was very different, tempered as it was by proximity, for we had walked right up to it closest face, and by the knowledge and shared histories of the locals in our group. For the latter, the encounter was challenging emotionally, tears flowed almost as fast as the glacier seemed to be melting, for the dramatic changes that have occurred in this valley once filled with ice and snow. Even in the last few years these tongues have retreated quite noticeably like a turtle pulling its head back into its shell. While we stood there, large chunks of ice caved off and new rivulets of meltwater were created. And at the same time memory and pictures show us quite clearly how much the carapace of this great frozen turtle has shrunk. Tons of ice and water have been shed such that the body of the glacier no longer stands high above the rim of its carved bowl being restrained by mountains of stones, but slumps mollified, easily contained.

No longer does it gleam blue and silver against the sky of grey the way a pearl drop of water is held skyward by the leafy fingers of an arctic lupine. Now it is cowed and dirty, a victim of the violence of changing climate and rising global temperatures. Damage done, in part as a result of my own desire to travel, to live comfortably in modern splendour, and to have the opportunity to walk with, on and under, ironically, glaciers such as these.

Desired by a belief that these encounters with the living wild and ancient beings enliven my own existence and expand the range of my own possibilities as human. So, what does one do with this knowledge, my own love and care is dangerous and harmful. This imposing being those that sparkles and exudes grandeur, that makes my heart sing and makes me laugh out loud, that I feel kinship with, and without whom I would be lesser than I am … is in fact dying. A violent death that is a result of my love and inertia, my actions and behaviours, the assertion of my own freedom and desire to flourish. Living and dying, loving and killing, tears and laughter, perpetrator and victim … frozen in the paradox of disappearing period ice.

Journal Entry, Part 3—Sean Blenkinsop

Fig. 3.12 "Frozen in Paradox" 35 mm Film Image: Photographer Karen Malone

And so, friend-visitor, as the sun sets on this day and these shared stories bounce around inside you, we hope that you have enjoyed, been touched and challenged by, the encounters. This is not an end so much as the second act in this larger play which will continue with the tomorrow's tales drawn from our group's meanderings to a high mountain cabin. The place that shaped Arne Næss. But first, some thoughts as to what was drawing/affecting/influencing/dissonating us, the curators, in these displayed works.

For us, these emotions of grief, sadness, horror, and pain we gathered in this idea of weeping are indicators of complexity but also of connection. Despite best attempts to not know relation, to not feel connected, to separate from the larger world, and to create structures and understandings that position the other as lesser than, there is still empathy, a capacity to touch and be touched, and the spark (often tiny) of inter-relatedness. Feeling and relating to the other remains a capacity. This spark then can be nurtured and built into flames as one finds ways to stay with the sadness, to build relations, to find calm in paradox, to see through opacity, to put in the time and practice (Jickling et al., 2018), and to unsettle the colonial and cultural ideas that have allowed the violences to continue unabated. As this chapter's stories notice, these emotions can act as a wake-up call.

Requests for attention cutting through the deep grey asking us to dissonate and undo comfortable/troublesome ways of being. Pushing us to turn a critical lens to the whys and wherefores of colonisation and parallel oppressions (Malone, 2023), to humbly reach out and seek/rebuild relations, and to find the courage to see oneself as having the potential to be both victim and perpetrator and to contain the latter in order to erase the former. For us, on this day, the work was to listen. To quieten the human voice and find ways to hear the colonised other/s. (Blenkinsop et al., 2017).

And yet, too much weeping and that flame is extinguished. For it is possible to wallow inert in despair or to make this yet another "all about me affair". "Oh, I am so sad the world is melting, woe is me". Then the focus shifts to responding to that human sadness while no attention is turned to those others whose suffering continues. Narcissism and anthropocentrism take over and the possibility for change—real, unsettling, eco-socially just (Blenkinsop et al., 2021), mutually beneficial flourishing (Blenkinsop & Morse, 2017) change—evaporates.

So, witness and weep; these are necessary. But don't be trapped for if our relations are to change, we must hold these insights scattered throughout this day held in ice. These tiny moments of wakening which challenge us to get to work changing ourselves and our relations or all that noticing, and pain will have been for nought. Sun sets, our storytellers retire, quiet sobs are heard as we await the tales that tomorrow brings. Stories of waking and working and a day in a cabin stuffed with objects igniting rememberings and tales yet still to come.

References

Blenkinsop, S., & Morse, M. (2017). Saying yes to life: The search for the rebel teacher. In R. Jickling & S. Sterling (Eds.), *Post-sustainability and environmental education: Remaking education for the future* (pp. 49–61). Palgrave Macmillan.

Blenkinsop, S., Affifi, R., Piersol, L., & Derby, M. (2017). Shut-up and listen: Implications and possibilities of Albert Memmi's characteristics of colonization upon the "Natural World." *Studies in Philosophy and Education, 36*(3), 348–365.

Blenkinsop, S., Fettes, M., Cole, L., Sage, S., Humphreys, C., & Chang, D. (2021). *Final report. Living within the Earth's carrying capacity: Towards and education for eco-social-cultural change.* A SSHRC funded knowledge synthesis

Jickling, B., Blenkinsop, S., Timmerman, N., & Sitka Sage, M. (Eds.). (2018). *Wild pedagogies: Touchstones for re-negotiating education and the environment in the Anthropocene.* Palgrave Macmillan.

Kimmerer, R. W. (2013). *Braiding sweetgrass: Indigenous wisdom, scientific knowledge, and the teachings of plants.* Milkweed Editions.

Malone, K. (2023). *Indigenising Earth Knowledges: Rising from the Ashes,* Keynote Address, Unity Earth, Peace Week Conference, Online Plenary, 16–23 Sept 2023

Martineau, J. (2014). Fugitive indigeneity: Reclaiming the terrain of decolonial struggle through Indigenous art. *Decolonization: Indigeneity, Education, and Society, 3*(1), 1–12.

Simpson, L. B. (2017). *As we have always done: Indigenous freedom through radical resistance.* University of Minnesota Press.

CHAPTER 4

Awakening-with Arne

On the third day of the gathering, we shifted our ramblings east of Finse, by two train stops. Our intent was to visit Arne Næss's cabin, *Tvergastein*, in the mountains high above the tracks. Arne[1] was a preeminent Norwegian philosopher and, in his later years, the face of Nordic ecophilosophy. Tvergastein was the legendary place where he did much of his thinking and wrote many of his books—away from the noise of the university. He often said that it is difficult to think "small thoughts" in a place like this. We hoped that being present in this same place would imbue fresh perspectives, larger thoughts, and refresh our understandings of relationships such as those that he developed within this landscape. And we hoped that being present would allow unseen, unfelt, untold dimensions of Arne's mountain experiences to arise from the flatness of his words as they now exist on the singular plane of written pages.

While hiking across the high sæter—or mountain pasture—we could see Arne's cabin. It hugged a ledge on the flank of the mountain Hallingskarvet, the highest private cottage in Norway. Per Ingvar Haukeland,[2] Arne's friend and colleague, was our guide and storyteller. He encouraged us to look up occasionally; sometimes it appears that "Tvergastein is coming to greet you". Perhaps. But that might require a more solitary walk.

By the time we reached the cabin, the temperature had dropped, and mist had wrapped our progress in a thin, damp blanket. One thousand

K. Malone et al., *Wilding Ecologies, Walking-with Glacier*,
https://doi.org/10.1007/978-3-031-63182-5_4

five hundred metres of elevation might not seem high to some, but at 60° 5′ north latitude, the conditions are arctic. Cables cinching the cabin to rock foretold of hurricane force winds. They also held stories of roofs ripped off.

During our walk, we added extra layers of clothing; we bent to look at flowers; and we skirted around a dead lemming. As we entered the cabin, we saw vials of Arne's chemicals, in blue hues. Further inside, we were greeted by Arne's chair sitting behind a blue table and by a large window facing south. And on its sill sat a row of large stones.

Wild pedagogical meanderings have persistently led us to wonder how a place—given a chance—might assert itself as a co-teacher. Now we wonder, could a place also assert itself as a co-researcher? Or co-philosopher? Traditions of Nordic Ecophilosophy suggest that real places have always been part of the mix—that ecophilosophy has always rested on more than abstract texts (Karl Georg Høyer and Petter Næss, 2012). The real world sits at the heart of these traditions. It is important to us, too.

Entering Arne's cabin is to begin stepping beyond witnessing and weeping. Tvergastein was certainly a site of much ongoing witnessing. However, this was also a place of awakening and working. It was where a philosopher lived, was awakened by the agency of the place, and worked to embrace a life reflected in "simplicity of means and richness of ends" (Næss, 2005, p. 351). Indeed, this is where Arne developed his personal philosophy, Ecosophy T—T suggesting Tvergastein. To enter this place was an opportunity to catch the spoor of a life lived, a place inhabited, and an inhabiting place. Here, theory and practice are one.

We are joined by two new voices on this novella's third day. You will hear Per Ingvar Haukeland tell stories about Arne at Tvergastein. You will also hear snippets of Arne's own voice drawn mostly from his essay, *An example of a Place: Tvergastein,* (2005) and from the movie *Crossing the Stones* (Horne, 1992), where he speaks about this place. The stories that follow continue a roughly conversational pattern as the voices move each other along, aligned from the left margins and then the indented margins of the pages. This gallery is curated through the agency of artefacts encountered along the way: Arne's chair, his table, a row of stones on his windowsill, and Tvergastein itself.

Fig. 4.1 "Arne's Cabin" Digital Image: Photographer Karen Malone

Sitting in Arne's Chair

In the last few months of Arne's life, I had the idea to take him back to Tvergastein. He was living in an old person's home, at 95 years. He spent his days slumped in a wheelchair. Fortunately, there was a sympathetic doctor, "what harm would there be if he died on such an excursion?" A snow scooter was arranged, and Arne was placed on it (Per Ingvar Haukeland,[3] personal communication, August 28, 2019).

Wrapped in fog . . .
I try to remember what period was.

Estella Kuchta

Gray eats everything I was going to say.

Stacy Boe Miller

Fig. 4.2 "Arnes Chair" Digital Image: Photographer Karen Malone

When they arrived, he was carried in and set to rest on his chair by the window. His face began to brighten. Suddenly he got from his chair and walked over to check the fire. He was also going to get more firewood! And he was alive again. Such was his relationship with this place. (Per Ingvar Haukeland, personal communication, August 28, 2019).

In this chair
by the window one last time.
Check the fire, get the wood, be the story
written by this place
one last time.

Bob Jickling

In front of the chair sits the oldest table in the cabin. Its blue paint is scarcely marked. To the left is a large window. Stones are placed on its sill.

On a clear day Arne could see a broad swath of Southern Norway. On some days, he saw the towers and wires carrying hydroelectric power to Oslo.

Perhaps on such days, he and his companions would playfully compete to see who could boil a cup of water by using the precise amount of fuel required—no more and no less.

What does a gallon of boiling water mean in the cities? (Arne Næss, 2005, p. 342).

On countless other days he would see grey mist, pelting rain, or just snow and whiteness in everything. In winter spindrift seeped through cracks in the wall and drifted into piles in cold rooms behind him. Sitting in his chair, he listened to the wind (Næss, 2005). And, perhaps, he pondered the stones resting on the windowsill.

Fig. 4.3 Sitting in Arne's Chair, Pinhole Camera, Photographer, Marcus Morse

What stuck with me from that day at Tvergastein was the lived experience of sitting in Arne's chair, searching across the landscape, and imagining the constantly changing views. Per Ingvar recalled his understanding that elements of Arne's philosophy could be drawn from those views, and that many such elements were rooted in the place itself. Ideas such as *pluralism*—experienced through constantly changing views, seasons, and weather; *possibilism*—the sensation of possibility and hope when looking across the mountains, and later moving in that same landscape; *pragmatism*, the need to act on one's belief or in the defence of a place; and *positivity*—the deep joy of life, that might be experienced through a walk in the areas surrounding Tvergastein.

Marcus Morse

The table interests me. It is most likely the very table that gave place to an epiphany that occurred more that 70 years ago. Arne, the ever-curious amateur researcher, often practised his own version of Tvergastein chemistry. At this worktable he created solutions that produced colours disclosing themselves in smooth waves, and never repeated variety. For him, the most elementary chemical processes revealed a fascinating world.

Bob Jickling

Arne recounts this experience:

I looked through an old-fashioned microscope at the dramatic meeting of two drops of different chemicals. A flea jumped from a lemming that was strolling along the table and landed in the middle of the acid chemicals. To save it was impossible. It took many minutes for the flea to die. Its movements were dreadfully expressive. What I felt was, naturally, a painful compassion and empathy. But the empathy was *not* basic. What *was* basic was the process of identification, that "I see myself in the flea". If I was alienated from the flea, not seeing intuitively anything resembling myself, the death struggle would have left me indifferent. So, there must be identification in order for there to be compassion. (Arne Næss, 1988, p. 22).

Blue Table

That the flea incident happened here, at this table, is important. While witnessing the table ourselves, we felt the porousness of Arne's cabin; it was drafty; the barrier between inside and out was thin. Arne was famous for keeping friendly company with the mice who readily slipped in and out. By now our group was also familiar with the abundance of lemmings in Norwegian mountains. *Of course,* they would have been inside *this* cabin too, and scampering across the table. It was *this table*, *this flea* and *this cabin* where Arne witnessed interspecies identification.

For *this* philosopher, *this* was a time of witnessing, and in his "painful compassion" perhaps a kind of weeping, too. And then, there was the awakening, as if being struck by a sudden bolt of *thisness* he realized a profound new understanding of relationship between identification, empathy, and solidarity with *this* flea (Zwicky, 2003). It was a moment of insight that fundamentally shaped this work.

In hindsight we also see an emergent example of flea as co-teacher, and co-researcher.

Rock climbing lived large at Tvergastein. Sometimes, Arne climbed several times a day during the long light of summertime. He reveled in the simple joys of rhythm and movement, and the appreciation of lichens, animals, and the sky. He was fond of the plants, particularly the *Saxifragia oppositifolia* (Purple mountain saxifrage). This "tiny" flower was the first to

bloom in the spring, high on the cliff faces where nearly twenty-four-hour-a-day sunshine warmed the rocks. His ecological and geographical observations were on the same plane as the climbing (Næss, 2005).
Per Ingvar, himself sitting in Arne's chair, reached to the windowsill and took one of the resting stones. Cradling it in his hands, he told another story about Arne's life as a climber.

> It began, as we would expect, on vertical routes. When an ageing Arne felt that he could no longer do this, he took up traversing—or climbing around the cliffs, at the same level, but at a low height.
>
> When Arne was no longer able to traverse the cliffs, he started climbing large boulders around his cabin.
>
> In time, Arne could no longer climb the boulders. Instead, he came inside and started climbing his chair—up and down from the floor to the seat.
>
> When he was no longer able to scale the chair, he reached for one of the stones from the windowsill. (Per Ingvar Haukeland, personal communication, August 28, 2019).

While it isn't clear if he was climbing in his mind, or whether he just needed the feel of the rock, these stones were a sensuous link with the physical world of Tvergastein.

Fig. 4.4 "Arne's Stones" Digital Image: Photographer Karen Malone

The Stones

Arne embraced his physical and sensuous being. He traversed the contours of the landscape that was Tvergastein—he walked, he skied, he climbed, and he played. He touched the rocks, he felt the warmth of sunlight for nearly twenty-four hours in a day, he bent to look closely at the early blooming *Saxifragia oppositifolia*. He stepped carefully around the most vulnerable plants growing in and around the rocks that were Tvergastein. And then, as his mind failed and his mental capacities dimmed, he reached out to his stones. He touched them and held them close. They beckoned him when grey had eaten most everything else.[4] They called him with a kind of knowing that exists beyond logic and language.

He still enjoyed a relationship with these bits of place, as if they and he were as one.

They gave him life. We wonder if we might even say that they were his *life*.

Arne suggests that Tvergastein is the name of his cabin, yet he cannot separate the dwelling from the place in which it is situated. He cannot separate cabin activities from outside activities. They are one to him. In his own terms, the high mountain surroundings and the cabin are the greater gestalt that is Tvergastein.

While staying at Tvergastein, one's attitudes change, and one's personality changes, at least temporarily. After one week there is a notable difference; three weeks—that is a very good stay... you are genuinely *there* (*Arne Næss*, 2005, *p. 344*).

For Arne, with time and practice, place can become *Place*—when the borders between oneself and place dissolve. When place becomes part of oneself—one's identity—you have *Place*.

Only Philosophies that impose sharp subject-object dualism try to trace a border between the self and "its" geographical surroundings... it is remarkable how a Place, even when is it is uninhabited most of the year, largely determines one's attitudes, one's likes and dislikes, and one's general outlook. One is caught up with the Place, hopefully with good consequences... (Arne Næss, 2005, p. 352).

I think, how could this place not be the co-author of Arne's philosophical work and his co-researcher.

Bob Jickling

In exiting Arne's cabin there was one final glimpse into his life at Tvergastein. Friends had placed a now weathered note in the window, addressed "Dear Hiker!":

Included in this note: ...Arne Næss has spent altogether about 14 years of his life in this place and wrote many of his books here... we hope the view can inspire us all to be as open to a life more in tune with nature. Arne Næss was especially fond of the vulnerable vegetation around this cabin, so we hope that you will admire and respect all life while you are here and tread lightly, mostly on the stones.

On the way down Arne cried. It was the only time I saw him do so. (Per Ingvar Haukeland, personal communication, August 28, 2019).

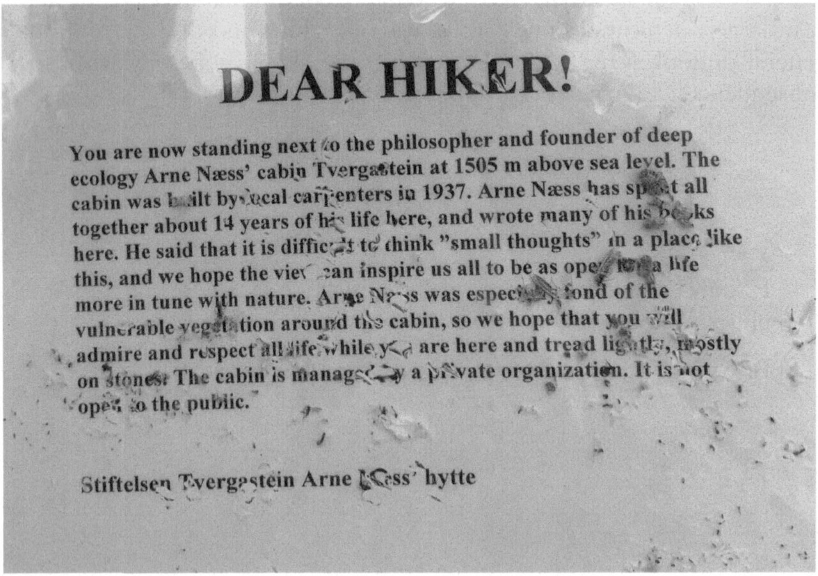

Fig. 4.5 Message to Hikers, Photographer Bob Jickling

Tvergastein

Arne's years at Tvergastein involved a personal journey from a small place clinging to the edge of a high mountain to a greatly expanded presence that he called *Place*—inseparable from his own identity. He traversed this journey with agents of this place—the flea, the *Saxifragia oppositifolia*, the mice, the lemmings, the boulders, the stones, the weather, and the endless other particularities of Tvergastein. He understood that in relationships of Place, the wellness of each agent is bound to the wellness of all, especially the vulnerable vegetation around his cabin.

We can read this Place in his return to "life:" in his bones that lifted him from his chair to get firewood, in his hands that were comforted by the touching of stones, in his tears of departure.

We can sense the presence of the table and catch a glimpse of an awakening that leapt from this place.

We can feel, in our cold bones, the work that Arne did in this place—to dwell with simple means and rich ends.

At Tvergastein, we are invited to wonder about our own awakening and our own work.

In the end we return to where we began:

Arne's Chair

Sitting in Arne's chair is to be held by a touchstone. When mentorship has been conducted mostly through words on flat pages, returning to Tvergastein is to touch their source. Sitting in this chair invites his words to fill the cabin, to be inflated by the presence of place, to rise like sourdough. It is a chance to glimpse, again, into the forces, feelings, and relationships that were Arne Næss.

Moving a short way down the train line to Arne's cabin broadened our landscape. It was an extension of our Finse experience. The mountains were familiar, as were the vegetation and the grey mist. Yet the place was unique; it was *the* place with which Arne Næss most identified; in an important sense it *was* him. In turn, this place gave breath to the tradition of Nordic Ecophilosophy. There were other ecophilosophers that were as rugged as the gneiss we walked upon, and as rooted to the landscape as Arne. They, too, filled this emerging philosophy with kith and wonder. Still, Arne's spirit was nearby and our short visit to Tvergastein invited us to ponder a network of trails that led beyond witnessing and weeping—a curriculum, you could say, of awakening and working. A little more background might be helpful.

Arne's affection for this landscape reached back to his childhood. His father died a year after he was born but left his family a small cottage above treeline at Ustaoset—the point of departure for our climb to Tvergastein.

From the time he was a small boy, Arne lived in this cottage in the summer and at Easter with his mother, sister, and two brothers. By his own account, he inexplicably found freedom above this treeline. By the time he was ten years old he had a dream of having a tiny cottage on this mountain and a desire to live there most of the time. He had already begun to see the mountain, Hallingskarvet, as a living being. Then, at the age of 25, in 1937, he began choosing a place for his cabin high on this mountain, and the dream began to take shape. (Horne, 1992).

By this time, Arne already had what he called a kind of practical philosophy—he had begun to build a cabin where he felt free to think and where he could live a simple life. Tvergastein was destined to become the place where he felt he belonged.

However, two years later, in 1939, he was made Professor of Philosophy at the University of Oslo. His immediate sensibility recoiled at the thought of being coerced to live on a campus, or to be present on a campus at all—except on rare occasions. For him, being a philosopher was very different from being a professor of philosophy. He recounts declaring that he would be on the campus Tuesday from 6 o'clock until Wednesday 3 o'clock, and nothing else. For him, a philosopher needed to work somewhere less hostile to the task. They needed to be where there is some kind of action in a context relevant to their work—"up in the mountains, or in some kind of conflict zone, or something" (Horne, 1992, 12:20:12:25).

For Arne, experience of place can put you on edge, disrupt assumptions, and provoke imagined possibilities.

Here Arne is naming a problem. For him, socially constructed norms about what a philosopher should do, are not necessarily consistent with the role of a philosopher. The task here, of course, is not to emulate Arne Næss, or even to determine if he is correct, rather it is to ask what work can this story do? For example: What social norms and norms of formal education constrain the work of wilding ecologies? Can this work be contained in abstract ideas, logic, and within the walls of schools and universities? To what degree are current norms of rationality socially constructed? Is more required? A possible first step in re-imagining ways forward, beyond our own witnessing and weeping, is to name the problems.

From the beginning, Arne had an expansive view of philosophy. He wasn't interested in small thoughts. He mused about the need for philosophers with a kind of old-fashioned sensibility of breadth and depth—and from these a willingness to act. "You need", he said, "people who are living according to their philosophy in the sense that their philosophy is relevant to what they decide every day" (Horne, 1992, 15:25:15:30). And this sensibility could be felt, sitting in Arne's chair, listening to Per Ingvar's stories, and then descending from Tvergastein.

Arne was once asked directly if asking deep questions—in the philosophical sense—was enough to challenge social norms and apparent truths? No, he declared, "you have to, if you can, get into some practical situation—you start a walk somewhere, do something together and then—bang—you have an example: 'How do you look at this?' Then you may make this person change the formulation of what they have seen as absolute truths" (Næss & Jickling, 2000, p. 51).

Here a pattern repeats itself; for Arne, physicality is an important access to knowing the world. He doesn't just "think" his way across a landscape, he feels his way. Acting isn't just a consequence of logical thinking, it is, indeed, another way of thinking and knowing, at times even reaching, outside socially constructed norms of rationality. At least those constructed in a western, modern, or European sense. And it seems that this knowing can function without words and logic. When the mind is wrapped in the fog of dementia, the body seemingly knows to put wood on the fire, derives sustenance from touching rocks, and sheds tears when leaving *Place*. In another sense incorporating bodily encounters in knowing can even be powerful enough to disrupt deeply held "truths", or create awakenings.

Of course, a playful imagination can cultivate opportunities, too. It was an unorthodox approach to chemistry in the company of lemmings and fleas that may have led to one of the most significant amongst Arne's awakenings. In an instant, looking through a microscope on that blue table, Arne identified with that flea. His recognition suddenly announced itself, not as a collection of logically linked parts; rather, it arrived in a moment as a whole understanding. Jan Zwicky (2019) calls these whole understandings gestalts that arise suddenly and affect us as sensuous beings in the world—as Næss was himself (1989). Zwicky considers these gestalt understandings to fall outside of socially constructed realms of logic and language.

These observations suggest that imagining and, indeed, enlivening an expanded version of rationality isn't an abstract task. It requires an ontological shift. It requires being in the world differently, and it may often require engaging in experiences that are held, felt, and disruptive (Jickling, 2017). Arne's witness of the flea in a bath of acidic chemicals allowed him to identify with the pain of this dying creature. And through this knowing experience, he was able to fundamentally re-imagine his relationship with place. In a sense, visiting Tvergastein was aiming to enable this kind of being-in-a-place. We wanted to feel what it was like to experience the mist and the cabin, and, indeed, sitting in Arne's chair. The Finse gathering was itself an experiment with felt and disrupting experiences, and wilding ways of knowing and being in place (e.g. Jickling & Morse, 2022; Jensen et al., 2022).

It is tempting to think that what we are calling "awakenings"—that is sudden bolts of insight—just appear as random experiences, as in lightening, never striking twice in the same place. However, they often require cultivation, opportunity, and openness (e.g. Jickling, 2017). Arne recommended walking, and doing things together, with other beings, outside. Who knows what awakenings might yet be rising from sitting in Arne's chair, rambling around Finse, or weeping with Marianne? Sometimes awakenings linger; sometimes they take longer.

Again, we ask how these stories might guide our work? How long does it take before we can *be there,* in a place? What is our work in questing *Place*—as individuals and as educators? Knowing in *Place*?

It is intriguing to wonder about other forms of knowing beyond social conventions, and those prioritised by schools and universities. Zwicky (1992, 2003) reminds us that our inclination towards logic, and its "rules of thought", may be in tension with an intuitive acceptance of suddenly revealed insights, and the acceptance of gestalts that resist analysis. They may also be in tension with a capacity to wonder at glaciers, spiders, ancient vegetation regurgitated, a long-dead flea, and a chair inside an old cabin. She adds,

> There is no series of steps we can implement to precipitate gestalts in all audiences. Real thinking does not always occur in words; it can decay under analysis; its processes are not always reportable. This means that real thinking is in some sense wild: it cannot be corralled or regulated. But it is also the only access humans have to the experience of insight, to moral and mathematical beauty, to ontological vision. (Zwicky, 2019, p. 95)

Wilding the ecologies of knowing can be risky. When rules of thought are loosened, we can see the world with fresh eyes, acknowledge feelings, and embrace physicality. We often fret when we let "rationality" off its leash. Yet, this is just a manifestation of a very controlled way of knowing. We can be tricked by our senses and by our intuitive impulses. We don't always get everything right and therein lies a danger. Yet, knowing, and hence learning, cannot be fully policed. In Zwicky's words, "where the danger lies, there too lies meaningful life" (2019, p. 95).

In this chapter we have pondered awakening and the work that it can inspire. We have begun to wonder about the heft of our experiences in the cabin of an individual so rooted in *Place*. We have speculated a little about the role played by this place, as a teacher and co-researcher. And, we have pondered the durability of socially constructed norms. In this section we are led to wonder, what can happen when humans wander outside norms in thought, imagination, and practice. And, to wonder again about what constitutes learning, knowing, being, and even self, culture, and community. For us, this is all part of the work of wilding ecologies.

We are challenged to ask: what is lost when thinking is limited to only thinking in words? What is lost when: we rely on a narrowly conceived notion of learning and rational logic; we prefer to teach students to see a world that is reduced to its constituent parts; and, we deny the physicality of learners and expect most formal learning—including formal lessons in philosophy—to be conducted indoors?

Can we wild our ecologies if we do not embrace more expansive inquiry that begins with an understanding that the wellness of each earthly being—every flea, saxifrage, rock, glacial erratic, and even the colour blue—is bound to the wellness of all other agents? Can we recognise that Gaia is writing her own story, and that we are barely listening? Is it reasonable to deprive ourselves of "access to the widest range of ideas, imaginations, and possibilities?" (Blenkinsop & Kuchta, 2024). We think not.

To end this chapter, we return to Arne, and to Tvergastein. As we leave his cabin we enter, again, the grey fog, the cold, and the damp. Glancing back one more time we read the old sign in the window addressed to us—The Hikers—and we wonder about the days swelling into months and years. Cumulatively, Arne lived here for 14 years. Yet, it wasn't until 1973 that he explicitly wrote about ecophilosophy. And, as witnessing implies an obligation to act, what triggered this awakening? And, we might fairly wonder what took so long?

Awakenings appear to come from nowhere, yet they cannot be guaranteed, predicted, or controlled; the moment of awakening can often be sudden and dramatic—such as the writhing death of the flee. But this incident, in turn, rested in a curious interest in amateur chemistry, a peculiar—even aimless—activity for a philosopher. Norwegians sometimes call this kind of wandering—physical and intellectual—fumbling and tumbling. Perhaps this is the point. Perhaps imagination needs the widest range of idea encounters in the word, and a kind of broad, sometimes aimless, attention.

In her classic book about writing, Brenda Ueland speculates that imagination needs, "… long, inefficient, happy idling, dawdling, and puttering. These people that are always briskly doing something and as busy as waltzing mice, they have little sharp, staccato ideas…. But they have no slow, big ideas" (Ueland, 1938/2010, p. 23). We can wonder about the fumbling and tumbling that we did, wandering across the sæter, witnessing the melting Midtdalsbreen, sitting in Arne's chair—and dormant awakenings that these experiences may, yet, have cultivated—with no promises.

We can wonder, anew about Arne's own fumbling and tumbling. We can dig a little more deeply and find that he was influenced by Rachel Carson's book *Silent Spring* (1962). And that his first harbinger of ecophilosophical writing was a 1965 activist plea to protect a mountainous area on the west coast of Norway, Innerdalen (Devall & Drengson, 2015). Was it Carson, a love of Innerdalen, itself, or maybe the emerging group of Nordic ecophilosophers that nudged and awakened Arne from witnessing to working?

But we also wonder, can we just be present for a bit? Can we dwell a little longer with the astonishment of being called to Midtdalsbreen—the middle arm of Hardangerjøkulen glacier? Or the magnificent rocks of the Finse sæter, or the bog cotton? Can we dwell a little longer with the privilege of entering, for a time, the place called Tvergastein? Can we hold our weeping a little longer—long enough to get past our own anthropocentric sense of loss of the shrinking glaciers, long enough to weep for the glaciers themselves, and the spiders that live on their receding edges? Can it not, for a time, be enough to just behold? This is not to say that we should abandon the creative and activist work demanded by our witnessing. Perhaps, this is another way of saying that witnessing is work too, and that deep witnessing takes time and unfettered attention and, as for Arne, can carry on for a lifetime.

We shiver a little. We pull our zippers higher and flip up the hoods of our rain jackets. Some don mittens and wool hats. Then after last glances at Tvergastein, we drop into the fog—and onto the work that is to be done.[5]

Fig. 4.6 "Leaving Tvergastein" Digital Image: Photographer Rebekka Sæter

Notes

1. This writing is inescapably intimate, as was our visit to Tvergastein. For this reason, we have chosen to use Næss's first name, Arne, throughout this piece—to reflect something of that closeness.
2. We are grateful for Per Ingvar Haukeland's kindness in agreeing to guide us through the terrain around Tvergastein, and important portions of Arne's work. We are also grateful for Tvergastein, where the Self that was Arne Næss could flourish, and together they could be our co-teachers.
3. All passages credited to Per Ingvar Haukeland were from notes taken and carefully hand recorded at the time of the storytelling. As such they are not precise verbatim quotations, but we feel that they accurately reflect the content and spirit of his remarks.

4. Here we extend the metaphor generated earlier in this chapter by Stacy Boe Miller.
5. This is a nod to Gary Snyder.

References

Blenkinsop, S., & Kuchta, E. (2024). *Ecologizing education: Nature-centred teaching for cultural change*. Cornell University Press.

Carson, R. (1962). *Silent spring*. Houghton Mifflin.

Devall, B., & Drengson, A. (2015). Preface. In H. Glasser & A. Drengson (Eds.), *The selected works of Arne Næss* (Vol. X, p. lvx–lvxxi). Springer.

Horne, J. (Director). (1992). *Crossing the Stones: A Portrait of Arne Næss* [Film]. Norwegian Broadcasting Corporation

Høyer, K. G., & Næss, P. (2012). Introductory perspectives, 2. In R. Baskar, K. G. Høyer, & P. Næss (Eds.), *Ecophilosophy in a world of crisis: Critical realism and the Nordic contributions* (pp. 1–8). London: Routledge.

Jensen, A., Presthus Heggen, M., Bob Jickling, B., & Blenkensop, S. (2022). Wild pedagogies for change. *Canadian Journal of Environmental Education, 25*, 5–12.

Jickling, B. (2017). Education revisited: Creating educational experiences that are held, felt, and disruptive. In B. Jickling & S. Sterling (Eds.), *Post-sustainability and environmental education: Remaking education for the future* (pp. 451–648). Palgrave Macmillan.

Jickling, B., & Morse, M. (2022). Experiments with lyric philosophy and the wilding of educational research. *Canadian Journal of Environmental Education., 25*, 13–36.

Næss, A. (1988). Self-realization: An ecological approach to being in the world. In J. Seed, J. Macy, P. Fleming, & A. Næss (Eds.), *Thinking like a mountain: Towards a council of all beings* (pp. 19–30). New Society Publishers.

Næss, A. (1989). *Ecology, community and lifestyle: Outline of an ecosophy,* trans. and revised. D. Rothenberg. Cambridge University Press.

Næss, A. (2005). An example of place: Tvergastein. In H. Glasser & A. Drengson (Eds.), *The selected works of Arne Næss* (Vol. X, pp. 339–359). Springer.

Næss, A., & Jickling, B. (2000). Deep ecology and education: A conversation with Arne Næss. *Canadian Journal of Environmental Education, 5*, 48–62.

Ueland, B. (2010). *If you want to write: A book about art, independence and spirit*. Wilder Publications. (Original work published 1938)

Zwicky, J. (1992). *Lyric philosophy*. University of Toronto Press.

Zwicky, J. (2003). *Wisdom and metaphor*. Gaspereau Press.

Zwicky, J. (2019). *The experience of meaning*. McGill-Queen's University Press.

CHAPTER 5

Dreaming-with Glacier

The Sun Sparkles Diamonds

Today, we are on alluvial spill, surrounded by a world of water. From where our little group sits, at the actively crumbling toe of a retreating glacier, we are immersed in the sounds, colours, forms, and creations of water. Silt filled streams under us gather crystal rivulets draining from the great ice carapace before crashing off down the boulder strewn, ice carved valley below. Ancient air wafts across our cheeks. Beyond, the sun sparkles diamonds across the surface of the darkening lake. And above, clouds are building in the west hinting, warning, of rain to come. Closer, we can smell the earth inflected moist air as our gaze is drawn into the exposed blue heart of the glacier, thousands of years old but now cracked open and melting fast, with no respite, in this age of climate warming. And around me, droplets of saltwater sliding down these faces that I have come to know and respect in our few days together. These tears are reminders of our ancient sea-going pasts and our troublesome present. We have just been told a story about the glacier by our host. A mountain of ice that has been part of their history since she was very young. A mountain of ice that is dying. A death that strikes each of us in our own way and that impacts this place in myriad ways.–**Karen Malone**

For many humans the story of glaciers in present times is focused on climate change headlines, melting ice, rising sea levels, ruination, and loss. Jackson, writing about this phenomenon, suggests "A glacier-ruins

K. Malone et al., *Wilding Ecologies, Walking-with Glacier*,
https://doi.org/10.1007/978-3-031-63182-5_5

narrative is understood as a narrative about glaciers that tends to overlook the existing state of a glacier and/or glacier systems and speaks instead to imagined states of loss" (2015, p. 479). They argue today a glacier has become not just a glacier—a lively icy terrain depicted though imagined and reimagined fiction and non-fiction stories by humans—but rather a barometer for measuring ice melting, climate change, the impending consequences of the Anthropocene. A barometer of fear and loss. It is a race against time, threatening livelihoods, water security, and flooded landscapes. Polar bears floating adrift on sea ice in rising oceans. This is the single story of glaciers as a metaphor for climate change and melting ice. Glacier existence is depicted in a simple formula: on a cold planet, glaciers exist; on a warming planet, glaciers disappear (Jackson, 2019). The human-glacier story has become entangled in this predicament. According to Jackson, what is absent from this narrative of ruination is "the entirety of the life of the glacier", in all its complexity (2019, 17). What is needed, then, is a broad re-wilding of glacier ecologies in richer and more diverse stories. And it is this what we have learned from Glacier and that we hope to contribute through this novella.

At the end of the Decameron the ten protagonists gather around a fountain filled with cool mountain water and prepare for their return to the city. They are leaving a place of safety, seemingly away from the plague and surrounded by the comforts of the natural world. They are returning to a more populous, and still disease-ridden, city. But return they do, because it is important to bring new perspectives back to the fray, and to do so with hope and energy. For Boccaccio there is a recognition that life is complex but that it helps to come at problems with a sense of munificence and maybe even love.

The same might be said for the Finse retreat and manifestations of wilding ecologies. This is a complex project that requires attention when returning from Glacier. At times the world returned to seems overwhelming. However, times of retreat can provide opportunities to gather new ideas, rethink old ideas in new ways, gather strength, dream of future work, and collect tools for the tasks at hand. In this chapter we, the authors, draw out, and weave together, themes that emerged from the previously curated chapters. At times, we call back to some partial reflections, and at times we offer some pedagogical prompts drawn from questions emerging from the wild pedagogical touchstones.[1]

Learning in the Anthropocene

The critical force of the Anthropocene has created space for persistent critique of the "humanist" paradigm. In disciplines such as education, this has allowed new conversations to emerge around human-dominated global change, human exceptionalism, ontology, and the nature/culture divide (Jickling et al., 2018; Lloro-Bidart, 2015; Malone et al. 2017). As a disrupting tool, the Anthropocene reveals that the scale and impact of ecological damage is unequal, unethical, and unjust (Malone, 2018). Entrenched assumptions rest upon a foundation of mastery and control—over bodies, over beings, over Earth's systems, over the stories being told, and over what counts as truth.

> Hardangerjøkulen: gives zero fuck about/the Anthropocene or some tone-deaf tribe/who thinks only itself well-versed. the glacier/ is lost in reverie again, composing life forms/& blue: it cares only for music—it *is* music.
>
> **Michael Dé Danann Datura**

Releasing control and undoing mastery is about changing the systemic fabrics of injustice. As such, wild pedagogues will likely find this loosening of control requires pushing a little further than just allowing learners extended freedom. It is this problematising of the human penchant for control that lies at the heart of the Wilding Ecologies project and Wild Pedagogies as a disrupting educational approach. To be "wilding" ecologies is not simply a theoretical endeavour. It has a theoretical, practical, and an educational one.

> How can we recognise human-centred habits, dominating impulses and urges to control and "manage" the natural world around us, and within our curricula? And how can we respond to these tendencies?

Reflecting on Wilding Ecologies

If our setting and Glacier's changing realities are key interlocutors, then several questions come to mind as we reflect on our experience: How did we story our time together? What unique ways emerged to consider theories, thoughts, modes of being, ideas, and voices collected on the mountain? And did we do justice to more-than-humans voices? We attempted to dream up methodologies, create encounters, and plan journeys into wilding possibilities, as writers, researchers, pedagogues, and creatives. And this thinking is entangled with ethics, ontologies, and epistemologies.

Feminist scholars espouse similar musings. They trouble ecologies imbued with conceptions of rational objectivity, patriarchal assumptions, (Merchant, 1983; Plumwood, 1993; Shiva, 1997), and culturally enforced gender types. Similarly, Indigenous scholars critique colonialism and think with expansive ecologies and cultural ways of knowing (Simpson, 2014; Tuck & Yang, 2012; Zembylas, 2017). Ecophilosophers challenge accepted norms of philosophical investigation and presentation (Næss, 1988; Zwicky, 1992; Turner, 1997), and ecocentric scholars argue that non-anthropocentrism is an imperative if we are to respond to the ecological crisis (Kopnina et al., 2018; Malone et al., 2020). Many of these scholars also have a presence in a posthumanist critique of humanist assumptions (Braidotti, 2019). And so, we see all of this as the work of wilding ecologies of being, knowing, and practising.

Questions about how to engage with emergent theoretical ecologies are significant. How can we tell stories of being-with Glacier without re-inscribing the injustices built into what are "accepted" as academic processes? In choosing the novella we are attempting to respond to these challenges. This form enables us to invite human and beyond-human collaborators to be with us in our walking, witnessing, and weeping. It allows us to de-centre the human voice and to lean into other ecologies: of being researcher, of being teacher, and of being humanly present in active and thoughtful relationship with the more-than-human. The novella form invites us to notice, to pay attention, to stay with the trouble of messy entangled worlds of living and of dying in damaged, icy landscapes (Haraway, 2016).

This novella offers another way of encountering and imagining novel research forms. We suggest that it offers a means to value, and be immersive in, more-than-human lives and places. It offers access to realms where human ways of knowing are diffuse, incomplete, entangled with being, and collective. And, the novella, as a form of wilding ecologies, disrupts ways of knowing by being affected, showing up, and recognising beingness.

Being with bodies in places, like Somerville (1999) does in her *Body/ landscape journals,* allows for yet more "possibilities to consider the politics of bodies and the ethics of being in country on landscapes, seascapes, and atmospheres" (Malone & Crinall, 2023, p. 4). For, as David Abram writes, being entangled with the more-than-human, and to bear witness, is to consider our bodies as place:

> Our body is altered and transformed as it moves through different lands…it is because the body is itself a kind of place—not a solid object but terrain through which things pass, and in which they sometimes settle and sediment (Abram, 2010, p. 230).

Similarly, Finse gathered a group of beings with bodies in places, most notably around Glacier. And this novella is an expansive container for gathering expressions of our collective entanglements and, indeed, our settling into place.

> Feeling into the world and its myriad more-than-human denizens can be exhausting work, particularly if done with the awareness that many are threatened and that places such as these exist, tenuously, on borrowed time.
>
> **Journal Entry—Sean Blenkinsop**

Yet, as we deepen our exploration of bodies in places, our encounter with Arne Næss' Tvergestein was profound. To literally sit in Arne's chair, and move about in his cabin, allowed our group an opportunity to settle and sediment for a little bit in his *Place.* It also allowed us to glimpse into the entanglements of Arne's own being and work. But, perhaps more importantly, our bodily presence allowed us to feel a small sample of his transformation from a being-in-a-place, to a being-as-place. It extended an opportunity to lift Arne's written words off their pages and to offer understandings that reached beyond just words and logic. And it extended an opportunity to understand how Arne and Tvergestein could be the being-as-place that he called *Place.* The novella again provided a form to experiment with expression of such profound experiences that reach beyond the boundaries of conventional academic work, and conventional words and logic.

The Finse gathering provided an opportunity to embrace shared voices across disciplines and beyond. It acknowledged that we are in a time of renewed calls for trans-disciplinarity and boundary crossings, and that by engaging in this work we can offer opportunities for wilding ecologies. Yet, there is an acknowledgment also that humanities and social sciences are prone to polysyllabic words—code, we think, for an inability to communicate with neighbouring academics, and certainly not an interested public. The point here is that these kinds of border crossings don't come easily; we must learn how to do them—and put them into practice.

Too often academia appears to be preoccupied with a manic quest to be new and innovative—to be first. Many things are lost, including an ability to communicate outside of our fields of expertise. John Ralston Saul (1995) argues academic disciplines, such as philosophy for example, have often absented themselves from public debate on important issues. The reason he suggests is many academics "are caught up in the complexities of philosophical professionalism—a world of narrow specializations and impenetrable dialect. A corporation of philosophy" (p. 161–162). The result is that they have left public debate wide open and vulnerable to more cynical forces. We can argue about the minutiae of his argument, but if we believe it is important to cross disciplinary borders, and embrace conceptual diversity, and if we believe we can contribute to the public discussion about the emerging Anthropocene, then we must take Saul's concerns seriously. We must listen. We must change our ethos and learn how to do our work differently. And so, we offer some conceptual possibilities that emerged from our time together.

> What did I do today that required learners to be sensually present in their learning? To encounter the other, to feel, care, respect, and to notice the more-than-human world?
>
> Where are my old habits limiting possibilities?

Relatedness and Entanglements

Many encounters with Glacier, described in the stories of this novella, reveal discoveries of entangled and related lives. They remind us we are not alone as individuals or as humans, and that what we do, and how we do it, affects others. They also challenge us, as viewers, witnesses, and researchers, to comprehend the depth of our shared collective grieving.

> Now, walking alongside this mountain artery, this ceaseless maelstrom of riffles and eddies, pushes at my dam of emotion. What grief am I holding?
>
> **Lee Beavington**

We heard stories of reconnecting and working to overcome individual separations from the natural environment while at the same time feeling a paradoxical sense of kinship and alienation, partner, and antagonist. We sometimes wonder if we are just grieving for our own losses. Or, have we sat with the grief long enough to identify with the other agents in the landscape, as Arne did with the flea. Just as our grief for Glacier and Spider emerged as consequences of the encounters, they bring attention to a "re-focusing on the past, present, and future entanglements of human and more-than-human lives and fates" (Gans et al., 2017, p. 67).

The more-than-human world is an entanglement in which humans dwell, inextricably tied to a host of many. However, much of western modern humanity barely seems to have noticed that this world is writing a new planetary geostory (Latour, 2014). Many scholars suggest that the naming of this geostory, the Anthropocene, compels educators to acknowledge our fate as shared with the fate of others. These deep ancestral relations affect and influence our shared sense of wellness—each one depending on the wellness of all others—as often appeared through representations of witnessing and weeping while walking-with Glacier. However, "we-are-all-one" entanglements can be problematic.

> Although we were heading towards the same place, maybe our thoughts were not?
>
> **Marianne's story**

Ahmed (2004), and others, argue that differentiation is endemic, as revealed in the politics of inequity. This work often tells the story of distinctions between those we identify with and those we view as other. And Ahmed reminds us that even the politics of grief "constitutes some others as legitimate objects of emotion" (Ahmed, 2004, p. 191) and others less so.

These distinctions often convey the perceptions of some other-than and less-than beings, rather than wondrous beings that are all part of the necessary diversity of kinship. Not only do humans often distance themselves from other species as subjects of legitimate emotion and concern, but they also position other human lives as less important. These and other more-than-human kin are some of the missing voices of the Anthropocene.

> What are my teaching habits that can distance, background, undervalue, or denigrate the natural world?

Temporality and Imaginary Futures

While sitting quietly on the edge of this glacier, hearing the raging water flowing deep within, time became an intriguing challenge for many of our storytellers. Scientists estimate that the ice at Hardangerjøkulen Glacier is about 4,000 years old, yet linear time, fractured time, and time measured on a human scale limit our capacity to understand deeper time. Being enmeshed in worldly encounters not only appeared to provoke ontological shifts, but also provided space for immersion in different, possibly "wilder", entanglements of non-linear ecologies of time. What are the possibilities that might arise for the human teacher and learner if the scale upon which learning can happen expands? Are different spheres for knowing and being opened if we learn in shifting temporal modes?

> As I stared over Finse, I felt the anguish of that ageless being across the lake. I tried to imagine its disappearance. Tried to visualize the magnitude of melting. And with the weight of this knowledge, we wandered south towards the retreating ice.
> **Erika Kazi**

> How are we, together, able to step out of linear time of learning places and work in different ways? And to recognise that some will need more time than others?

Wilding notions of time also influences how we can come to be with global challenges. It is long-term thinking that is needed as a critical way to rekindle humankind's imaginary as we grapple with climate change. Ehrlich and Ehrlich (2013) invite us to nurture "foresight intelligence", that is, the ability to look ahead and act for the good of future beings even if short term costs are involved for us. Rather than quick-fix scientific and technological responses, "foresight intelligence" encourages us to consider moral and ethical concerns about "what is the right thing to do?" and "what are needs of future generations?" It encourages cautiousness, taking time to better understand the ramifications of the "new" (Paulsen, 2021).

Post-anthropocentricism is a way of thinking that disrupts and refuses to support human exceptionalism. It rejects the human pretence of sole mastery and control; we are not exceptional, and we are not superior to the entities who live, act, and die alongside us. At a global level, naming the current era the Anthropocene recognises the irreversible activities and impacts that particular humans have had on the planet. Capitalism, western thinking, individualism, over-consumption, and the human desire to control ecological resources are driving an ecological crisis. Yet paradoxically, as we realise the ecological devastation of human doings, we also realise that a viable response to these crises is beyond human control.

(Un)common Communities

The Finse gathering was an opportunity to embrace shared voices across disciplines and beyond. In a time of renewed calls for boundary crossings, through a range of alliances, working together offered opportunities for wilding ecologies. Indeed, storytellers often situated their contributions to the novella in a context of being *together*, particularly in relation to conversations about change.

> We bend ourselves / to the sun when we see / the sun. We eat / what the morning gives, / we point, / for you, / to where the wind is going.
>
> **Stacy Boe Miller**

Change is not easy, substantial change is even harder, and less likely to be successful if done individually. For we live in, are shaped by, and are always responding to, communities. We must do this work together. Braidotti (2019) suggests that in these times when we find ourselves caught between stasis and despair, there are two options: surrender, or reinvent ourselves. But, in community and in entangled relations, we cannot reinvent ourselves alone. As Braidotti asserts, we "need a people, a community, and an assemblage: 'we'—this complex multiplicity—cannot survive or act alone" (p. 18). The labour of educational transformation must happen in, with, and through community.

Arne Næss, in considering the difficulties of building community, notes: "we tend to specialize. With this specialization there is a tendency to feel opposites instead of feeling the complexity of the relations and complementarity" (Næss & Jickling, 2000, p. 50). He also suggests there are important convergences between the Deep Ecology approach, the eco-social justice movements, and ecofeminism. And, with a note of grace uncommon in academia, he says: "we are so grateful to work with these ideas" (p. 50).

With a nod to how we might learn to do things differently, Næss recalls his experiences as a young scholar amongst the Vienna Circle of philosophers. Although they were very different personalities, they were constantly helping each other with invitations to try things differently and to look for new meanings. This ethos constituted what Næss called a kind of "research attitude" (Næss & Jickling, 2000, p. 51). This is an attitude, we think, that can be extended to how mixed or uncommon communities can be together. Such collaborative shifts in research ethos, reflecting generosity in scholarship (Russell, 2006), would be an important move towards growing, working in, and building the kinds of community that could generate significant change.

Like the Critical Realists and the Nordic ecophilosophers who shared conversations and sought convergences (Bhaskar, Høyer & Næss, 2012), the gathering in Finse and, by extension, this novella, seeks to reveal convergences. Through convergences we collectively probe ways of being, knowing, and valuing that are more diverse, robust, polyvocal, and inclusive. Such convergences demand an "unlearning" of the western modernist human story, a movement away from anthropocentric ways of being in, valuing, and knowing the world. And they require a release of control and an ongoing disruption of static and the limiting ecologies of the status quo.

> How may my communities encourage one another to depart from the status quo? How do we encourage and challenge each other, allow for mistakes, and rebound from setbacks?

Disrupting and Expanding

Disruption provides the impetus for new learning in many of this novella's stories. Without disruptions, or some kind of psychic, emotional, or mental dissonance, there would be little opportunity for fresh insights to be considered. They open space for experiencing alternative ways of knowing and being and are central to the process of wilding ecologies of learning for individuals, communities, and even cultures.

> In this chair/by the window one last time./Check the fire, get the wood, be the story/written by this place/one last time.
>
> **Bob Jickling**

Sometimes these experiential disruptions can fundamentally transform individuals' core understandings of their relationships with, and in, the world (Jickling, 2017). Arne Næss, for example, in a sudden bolt of insight, discovered that he identified with a dying flea. For our wandering storytellers, tears pouring out over a melting glacier and spiders grazing on ancient plant matter become disrupters for the concept of time. In these cases, individuals experienced something that didn't make sense, or fit in with, presumed ways of being in the world. They experienced something that couldn't be placed within their pre-existing frames of understanding, or even something that ran contrary to what thought they knew.

At some level disruptive experiences are about understanding how one's knowledge is inadequate and incomplete. Yet, some dissonances also point towards larger and more culturally accepted truths. For example, they can upend understandings of linear human-scaled time, or even the colour blue. Cracks can appear in the ice of an assumed ontological hierarchy which places humans—some humans at least—at the top and centre of everything while at the same time positioning humans as exclusively independent of others. Taking this further, it is important to wonder how these disrupted concepts shed light on how knowledge is dispersed, care is enacted, including for the more-than-human, and how values, rights, and ethics can traverse the bounds of species.

Although such disruption of assumed truths, incomplete theories, and unfounded facts are potentially rich fodder for change, it is an incomplete process. To disrupt without offering something in response is to potentially damage the educational process. So, we add the idea of expansion to this conversation. Again, we draw from the novella's storytellers as we notice expressions of their experiences, their imaginations, and even how their own understandings are expanded. Experiences are fodder for the imagination. It is the concreteness of experience that can push the edges of imaginative range (Blenkinsop & Jickling, 2020). And one's understanding of things is nothing if not an amalgam of experience and imagination. So, while disruption can highlight the current limits of one's own thinking, imagination, ideas, and culture, expansion of opportunities for having diverse experiences—the "stuff" of learning—can generate as yet unimagined possibilities. Similarly, communities of learners working together within rich experiences can extend collective imaginaries (Young & Malone 2023).

> What new "stuff", experiences, and stories did I add to the mix? And, how are learners taking up, working with, and being changed by these diverse cultural tools?

Decentring and Embodying

Decentring and embodiment go to the heart of the Finse gathering. From the outset, the intention was to be in a place, listen to a place, and live into being partners with that place. Unsurprisingly, these experiences came to the fore as offerings from the novella. We partner decentring with embodiment because some novella storytellers highlight an important move beyond a centralised, controlled, and cognitive project of rationality, toward a more complex positioning of being in, and with, the world. And it was seemingly most often their bodies that allowed authors to witness, and weep, with Glacier, to recognise their relatedness, to come into place, and to embrace relationship and belonging. As Serres (1995) highlights, part of the modernist alienation is derived from the telling of a mythical story of human superiority and centredness. It is a story of control and exclusive ways of knowing, a tall tale, in fact, that requires a negation of other forms of knowing and being, particularly those that are relational and experiential. It is here that experiences of place offer possibilities for decentralised and embodied ways of being that are critical for wilding the ecologies of learning in education.

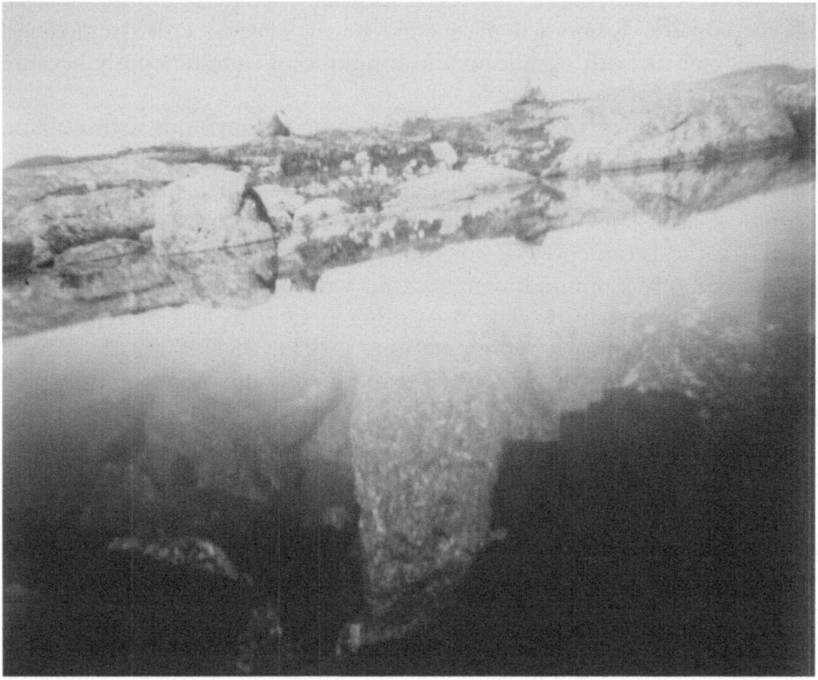

Fig. 5.1 Looking In: Pinhole camera, Photographers, Finse, Workshop Participants

> What have I done to accommodate experiences that exist beyond the capacity of language to fully describe and evaluate?

Decentring is an important way to acknowledge human-centrism and to actively release anthropocentric control, and other systems of exclusive power, at all levels of the educational project. This includes decentring the human teacher as the hub of learning and embodying practices that allow learners, their community, and the more-than-human to be co-teachers (Malone & Crinall, 2023). It recognises knowledge as diffuse, shared, and incomplete. Embodied decentring means staying with paradoxes—to see oneself as both significant and insignificant simultaneously because not all challenges are resolvable.

Embodiment has, in some spheres, become synonymous with walking as method/ology (Walkinglab.org, 2024). Here, researchers are interested in the corporeal and lived dimensions by which a walker moves through space. They consider the surfaces and textures of the ground as part of the production of emerging spaces. Sarah Pink, for example, locates the body as embodiment "within a wider ecology, allowing us to see it as an organism in relation to other organisms and its representations in relation to other representations" (2011: 354). Similarly, for the Finse participants, a wild ecology of embodiment emerged from the deep crevasses of Hardangerjøkulen Glacier.

Enacting and Decolonising

Decolonising is not explicitly referenced by the novella storytellers; however, it is implied throughout as a seeking of justice and undoing systems of the colonial project. The work of decolonising ourselves; our communities, knowledge systems, ways of being in the world, and practices involves wrestling with positions of power and, in turn, subjugation by external practices and internal assumptions, and by privileges gained by some and possibilities lost by others. Decolonising cultures is a journey that is never finished, and it is, in part, the work of wilding ecologies presented throughout this novella.

> Glacier talking back through stories, curated poems, and images.
> Glacier coughing, scoring, and ruminating in ancient landscapes.
> Glacier receding, vanishing, and haunting human imaginaries.
>
> **Karen Malone**

Decolonising enactments are spread throughout these pages. We hear them in personal projects that seek to position nature as co-teacher and co-researcher. We witness them in deep questioning and living into the answers. They arise in the presence of Glacier, spiders on melting edges, Arne's cabin, and through time spent with each other. We notice different forms of representation, as well, as authors come together to, write, dream, imagine, educate, and take ownership of ideas differently. We hear reverberations of these explorations played out as folks reach out, build communities, and deepen relationships with the more-than-human world. Working in this space is difficult; enacting new insights and practices means risking and making mistakes. However, the risks of inaction are no longer tenable.

> How can we help learners "lean into" difficult encounters with human privilege, alienation, and dominance?

Returning to Fragile Bodies

The term Anthropocene has been used throughout this novel as a lens to focus attention on the dangers of our epoch. Perhaps this term is too mild to capture the peril. For some, the Anthropo-unseen or Anthropo-obscene, are more accurate terms to capture the new assemblages of catastrophic times. In Finse, a past was traced onto the present allowing possibilities for a kind of "travel hopping" (Barad, 2007) through deep time, entangled in speculative imaginaries. As the assemblages of our fragile bodies reached out with threadlike tendrils encircling and clinging to imagined futures, we walked in single file across the glaciers melting, disintegrating edges:

> As humans reshape the landscape, we forget what was there before … our newly shaped and ruined landscapes become the new reality. Admiring one landscape and its biological entanglements often entails forgetting many others (Gan et al., 2017, p. G6).

Fig. 5.2 The Harvestman *Mitopus morio* and the wolf spider *Pardosa trailli*—Terresterial predators whose bodies contained ancient carbon released by Hardangerjøkulen glacier (*Source* Hågvar et al., (2016))

Harvestmen and Wolf Spider are spotted on the watery edges of Hardangerjøkulen glacier feeding on ancient carbon. Glaciers play a critical role in the global carbon cycle as they collect, produce, store, transform, and now in this time of carbon overload release carbon from its deeply held clutches. Freshly deposited sand and silt close to the edges of the receding Hardangerjøkulen glacier contains particles of thawing vegetation, this ancient carbon. Research suggests the ancient carbon released from the Hardangerjøkulen glacier, now traced inside lively bodies, originated from soil and vegetation in periods when the glacier was absent likely between 4000 and 8000 years ago.[2] Ancient carbon flowing from melting edges of Hardangerjøkulen leaves behind sustenance on the glacier foreland to support new inhabitants. Arthropods, beetles, and spiders, quickly colonise the newly exposed sludge. In rapid succession they develop food webs feeding on the decomposing plants and ancient carbon particles flowing from the glacier. As voyeurs of this moment of unfolding, the partiality of what we thought we could see or know, is revealed. As we encounter the retreating glacier, in the darkness of her grief and loss, there is the giving of new life, an ecosystem of super-pioneers. Old, ancient carbons, laid down before the settlers followed the reindeer into the newly formed meadows after the ice age, is slowly awakening from its slumber.

Grandmother glacier is rebirthing.

Karen Malone

Wild rivulets of ecologies run through this novella, shifting fragments of geologic time over a disintegrating, icy, and watery landscape. In turn, this novella works with flows of pedagogy, theory, and collective creative practice. The stories of walking with Glacier speak through photographs, prose, poetry, and provocations. Collectively, the gathering of experiences in this book explores what it means to be human and more-than-human in the context of glacial melt and shifting loss. We are changing our planet and, all the time, changing what it means to be human.

Major ecological changes, indeed, ecological crises, are situated across the nature/culture divide. The Anthropocene has the potential to disrupt a persistent "humanist" paradigm in disciplines such as education by allowing new conversations and wild(er) ecologies to emerge (Malone & Young, 2023). As a disruptive ontological tool, it reveals there is no homogenous/universal human species. Divisions between arts, humanities, and sciences compound the problems of that divide and inhibit the work needed to address present crises.

Through this novella we have sought to disrupt the moral axiology. As Drengson and Devall (2008) argue:

> When we care for our place and others, we come to identify with their needs and well-being, and we have a greatly enhanced and larger sense of community and interdependence... we naturally and spontaneously care for our place and seek to protect it. For this we do not need a moral axiology, a set of rules or enforcements held over us to force us to act. We are able to act beautifully with expansive grace, gratitude, and generosity. This brings us deep knowledge of others and a great sense of joy that can expand indefinitely. (p. lxviii)

Through this novella we seek also to offer possibilities for speculative becoming's cradled in experiences of walking-with Glacier. While walking, we weep and bear witness to the Anthropocene. And we care. Writing brings attention to the unseen; the stories "awaken ethical sensibility" (Rose, 2012, p. 139) and enlarge senses of community, place, and interdependence (Jickling et al., 2021). We have found through walking-with Glacier, as Jackson (2019) suggests, that the "Time has come for new ways of telling stories about Glaciers—and for listening to the stories Glaciers tell us" (p. 20). The novella has allowed us to explore the rich complexities of care and appreciation found not only in human communities, but also in more-than-human places of a wilder freer nature.

Our novella companions have departed. And we, the authors, have lingered a while to consider the work done in Finse. Throughout our retreat to the mountains we have, together, walked-with Glacier and dreamed-with Glacier as the beginning of a reawakening of love and enchantment for an uncertain future and a changing planet. We hope you, the reader, find resonance here within as you encounter the work of wilding your own ecologies.

Notes

1. See Appendix 1.
2. See Hågvar and Ohlson (2013), Hågvar et al. (2016).

References

Abram, D. (2010). *Becoming animal*. Taylor and Francis Publishing Group.

Ahmed, S. (2004). *The cultural politics of emotion*. Edinburgh University Press.

Barad, K. (2007). *Meeting the universe halfway*. Duke University Press.

Bhaskar, R., Høyer, K. G., & Næss, P. (Eds.). (2012). *Ecophilosophy in a world of crisis: Critical realism and the Nordic contributions*. Routledge.

Blenkinsop, S., & Jickling, B. (2020). Wilding teacher education: Responding to the Cries of Nature. *Canadian Journal of Environmental Education, 23*(1), 121–138.

Braidotti, R. (2019). *Posthuman knowledge*. Polity Press.

Drengson, A., & Devall, B. (Eds.). (2008). *The ecology of wisdom: Writings By Arne Næss*. Counterpoint Press.

Ehrlich, P., & Ehrlich, A. (2013). Can a collapse of global civilization be avoided? *Proceedings of the Royal Society B*. https://doi.org/10.1098/rspb.2012.2845

Gan, E., Tsing, A., Swanson, H., & Bubandt, N. (2017). Introduction: Haunted landscapes of the Anthropocene. In A. Tsing, H. Swanson, E. Gan, & N. Bubandt (Eds.), *Arts of living on a damaged planet* (pp. G1–G14). University of Minnesota Press.

Hågvar, S., & Ohlson, M. (2013). Ancient carbon from a melting glacier gives high 14C age in living pioneer invertebrates. *Scientific Reports, 3*, 2820.

Hågvar, S., Ohlson, M., & Brittain, J. E. (2016). A melting glacier feeds aquatic and terrestrial invertebrates with ancient carbon and supports early succession. *Arctic, Antarctic, and Alpine Research, 48*(3), 551–562.

Haraway, D. (2016). *Staying with the Trouble: Making Kin in the Chthulucene*. Durham, NC: Duke University Press.

Head, L. (2016). *Hope and grief in the anthropocene: Reconceptualising human-nature relations*. Routledge.

Jackson, M. (2015). Glaciers and climate change: Narratives of ruined futures. *Wires Climate Change, 6*, 479–492. https://doi.org/10.1002/wcc.351

Jackson, M. (2019). *The secret lives of glaciers*. Green Writers Press.

Jickling, B. (2017). Education revisited: Creating educational experiences that are held, felt, and disruptive. In Jickling, B. & Sterling, S. (Eds.) *Post-sustainability and environmental education: Remaking education for the future* (pp. 451–648). Palgrave Macmillan.

Jickling, B., Blenkinsop, S., Timmerman, N., & Sitka Sage, M. (Eds.). (2018). *Wild pedagogies: Touchstones for re-negotiating education and the environment in the Anthropocene*. Palgrave Macmillan.

Jickling, B., Lotz-Sisitka, H., Olvitt, L., O'Donoghue, R., Schudel, I., McGarry, D., & Niblett, B. (2021). *Environmental ethics: A sourcebook for educators*. SUN PReSS.

Kopnina, H., Haydn Washington, H., Taylor, B., & Piccolo, J. (2018). Anthropocentrism: More than Just a Misunderstood Problem. *Journal of Agricultural Environmental Ethics, 31*, 109–127.

Latour, B. (2014). Agency at the time of the Anthropocene. *New Literary History, 45*(1), 1–18.

Lloro-Bidart, T. (2015). A political ecology of education in/for the Anthropocene. *Environment and Society: Advances in Research, 6*, 128–148.

Malone, K. (2018). *Children in the Anthropocene*. Palgrave Macmillan Publishers.

Malone, K., & Crinall, S. (2023). Children as worlding but not only: Holding space for unknowing and undoing, unfolding and ongoing. *Children's Geographies*. https://doi.org/10.1080/14733285.2023.2219624

Malone, K., & Young, T. (2023). Retheorising environmental sustainability education for the Anthropocene. *Educational Philosophy and Theory Journal, 55*(11), 1200–1204.

Malone, K., Truong, S., & Gray, T. (Eds.). (2017). *Reimagining sustainability in precarious times*. Springer.

Malone, K., Logan, M., Siegel, L., Regalado, J., & Wade-Leeuwen, B. (2020). Shimmering with Deborah Rose: Posthuman theory-making with feminist ecophilosophers and social ecologists. *Australian Journal of Environmental Education, 36*(2), 129–145.

Merchant, C. (1983). *The Death of Nature: women, ecology, and the scientific revolution*. Harper Collins.

Næss, A. (1988). Self-realization: An ecological approach to being in the world. In J. Seed, J. Macy, P. Fleming, & A. Næss (Eds.), *Thinking like a mountain: Towards a council of all beings* (pp. 19–30). New Society Publishers.

Næss, A., & Jickling, B. (2000). Deep ecology and education: A conversation with Arne Næss. *Canadian Journal of Environmental Education, 5*, 48–62.

Paulsen, M. (2021). Cautiousness as a new pedagogical ideal in the Anthropocene. In M. Paulsen, G. H. Jacobsen, J. Garsdal, K. V. Brömssen, O. Koefoed, & K. B. Petersen (Eds.), *Rethinking education in light of global challenges: Scandinavian perspectives on culture, society and the Anthropocene* (pp. 220–235). Routledge.

Pink, S. (2011). From embodiment to emplacement: Re-thinking competing bodies, senses, and spatialities. *Sport, Education and Society, 16*(3), 343–355.

Plumwood, V. (1993). *Feminism and the Mastery of Nature* (1st ed.). Routledge.

Rose, D. B. (2012). Multispecies knots of ethical time. *Environmental Philosophy, 9*, 127–140.
Russell, C. (2006). Working across and with methodological difference in environmental education research. *Environmental Education Research, 12*(3–4), 403–412.
Saul, J. R. (1995). *The unconscious civilization.* Anasi.
Serres, M. (1995). *The natural contract* (trans. E. MacArthur & W. Paulson). University of Michigan Press.
Shiva, V. (1997). *Staying alive: women, ecology, and development.* Zed Books.
Simpson, L. B. (2014). Land as Pedagogy: Nishnaabeg Intelligence and Rebellious Transformation. *Decolonization: Indigeneity, Education & Society, 3*(3), 1–25.
Somerville, M. (1999). *Body/Landscape Journals.* SpinifexPress.
Tuck, E., & Yang, K. W. (2012). Decolonization is not a metaphor. *Decolonization: Indigeneity, Education & Society, 1*(1), 1–40.
Turner, J. (1997). *The abstract wild.* University of Arizona Press.
Walkinglab.org (accessed March 9, 2024). https://walkinglab.org/podcast/walkinglab-introduction-to-critical-walking-methodologies/
Young, T., & Malone, K. (2023). Reconfiguring environmental sustainability education by exploring past/present/future pedagogical openings with preservice teachers. *Teaching in Higher Education, 28*(5), 1077–1094.
Zembylas, M. (2017). The quest for cognitive justice: Towards a Pluriversal Human Rights Education. *Globalisation, Societies, and Education, 15*(4), 397–409.
Zwicky, J. (1992). *Lyric philosophy.* University of Toronto Press.

Appendix 1: Touchstones for Practice

The Wild Pedagogies *touchstones* introduced below are concepts intended to support the work of educators heuristically. They emerged from Wild Pedagogies gatherings and are voiced through the bringing together of participants and pxlace (entitled Crex Crex Collective). They recognise the difficulty in achieving sustained cultural change by providing ideas that can be held and returned to over and over—for potential reference, guidance, and support. Touchstones, then, are reminders of what we are trying to do, as provocations they help us return to new openings especially when daily demands become all consuming. They are intended to be revisited, refined, and reconsidered, but they can also stand as points for departure. These touchstones are drawn from experiments in practice and attempt to bring the more-than-human world actively into educational conversations.

Touchstone #1: Nature as Co-teacher

> We believe that education is richer, for all involved, if the natural world and the many denizens that co-constitute places, are actively engaged with, listened to, and taken seriously as part of the educative process. (Crex Crex Collective, 2018, p. 80)

At one level this touchstone seems quite easy to understand and even put into practice. The claim is that the natural world is a vibrant, active, agential place that is worth listening and attending to, and learning from.

K. Malone et al., *Wilding Ecologies, Walking-with Glacier*,
https://doi.org/10.1007/978-3-031-63182-5

Accepting this likely means that educators will spend more time outdoors and thus find different pedagogical possibilities appearing and new affordances being engaged. However, at another level, this touchstone has significant implications for what knowledge is and how learning happens. If nature becomes a co-teacher then the human, as the sole possessor, arbiter, and conveyor of knowledge is de-centred and learning becomes a shared project that is no longer ever complete or human-based.

With this discussion as background, educators might want to consider questions such as:

- How can I invite the natural world to be present as a co-teacher in my practice? To allow other teaching voices to be heard in their own ways?
- How have we been able to learn about, with, and from the more-than-human world?
- How can my class and I contribute to the flourishing of each other and those other beings that we live amongst?
- What are my teaching habits that can distance, background, undervalue, or denigrate the natural world?

Touchstone #2: Complexity, the Unknown, and Spontaneity

> We believe that education is richer for all involved, if there is room left for surprise. If no single teacher or learner can know all about anything, then there always remains the possibility for the unexpected connection to be made, the unplanned event to occur, and the simple explanation to become more complex. (Crex Crex Collective, 2018, p. 84)

This touchstone prioritises the unpredictable as it pushes back against the desire to control and contain. This allows for a diversity of voices and possibilities that are often marginalised or even lost in environments where the standardised, the measurable, the normal, and the definable are the focus. For educators this will involve some risk as the emergent tends to complexify situations and curriculum design can no longer focus on simply positing desired learning outcomes and then pushing students towards those chosen particulars. The suggestion here is that the world does not in fact work in such a clean, predictable, linear fashion and that something important is lost when we assume, thus creating educational practices built on that assumption, it does.

With this discussion as background, educators might want to consider questions such as:

- How did I embrace complexity in my teaching today? Was there room for the unknown, spontaneous, and unexpected to appear, and be taken seriously?
- How did I empower learners to embrace the complexity of knowledge and not reach for easy answers? Did they encounter the incomplete nature of knowledge today?
- Did I take risks today in moving away from attempts to control assumed ends? How can I continue that tomorrow?

Touchstone #3: Locating the Wild

> We believe that the wild can be found everywhere, but that this recognition and the work of finding the wild is not necessarily easy. The wild can be occluded, made hard to see, by cultural tools, by the colonial orientation of those doing the encountering, and, in urban spaces, by concrete itself. (Crex Crex Collective, 2018, p. 88)

This touchstone brings the critical into Wild Pedagogies as it cautions against both the cultural constraints of much of modern public education and the colonial orientations that modernity has towards the natural world and many peoples. Taking this seriously will challenge educators to think about their own privileges, including those related to the more-than-human world, and to be constantly aware of how the language and metaphors, the structures they work within, the tools they employ, and the ways they be either challenge or sustain problematic *status quos*. This touchstone also suggests that the wild can be located anywhere, in the rural and the urban, but also within any individual and that it can be, and often is, obfuscated by cultural and colonial overlay. As such, educators will be challenged to facilitate encounters with the wild that are able to name and respond in critical ways to those things that are seeking to stop learners from encountering their own freedoms.

With this discussion as background, educators might want to consider questions such as:

- How can I enable encounters with wild, self-willed, communities in the spaces that my students co-inhabit? And how can we ensure these encounters are acknowledged?
- How can we recognise human-centred habits, dominating impulses, and urges to control and "manage" the natural world around us, and within our curricula? And how can we respond to these tendencies?

How can we help students "lean into" difficult encounters with human privilege, alienation, and dominance?

Touchstone #4: Time and Practice

> We believe that building relationships with the natural world will, like any relationship, take time. We also believe that discipline and practice are essential to this process. (Crex Crex Collective, 2018, p. 92)

This touchstone focuses on two key discussions: process and practice. Both are ultimately interested in building and maintaining relationships with the natural world and particularly those denizens who live in immediate proximity to us. The first focus suggests that this is done through spending significant amounts of time in place, immersed in and listening to the world, but also that we reconsider how we conceptualise time and find ways to slow down, to change habits that keep us separate from others, to listen in different ways to our own bodies and the bodies around us, and to immerse ourselves in what some have called deep time. The second focus has a pair of meanings. In this case practice implies discipline, the work it will take to build rich relationships becomes like that which it takes to develop a meditative practice. Practice also implies the activity of one's pedagogy, how one teaches and the assumptions and habits that motivate that work. Wild Pedagogies assumes that building relationships as part of student learning will require a change in current practice.

With this discussion as background, educators might want to consider questions such as:

- Can I leave enough time and space in my teaching for engagement with the nearby natural spaces and more-than-human beings?

- How are we, together, able to step out of schools' linear time and work in different ways? And to recognise that some will need more time than others?
- Am I able to notice and support learners who are trying out new habits? Am I trying new practices myself, and reflecting on these attempts?
- How can I nurture my own immersion in places? And building relationships with these places and the beings encountered?

What opportunities do my students have to develop their intuition and other overlooked ways of knowing and understanding?

Touchstone #5: Socio-Cultural Change

> We believe that the way many humans currently exist on the planet needs changing, that this change is required at the cultural level, and that education has an important role to play in this project of cultural change. We also believe that education is always a political act, and we see wild pedagogues embracing the role of activists as thoughtfully as they can. (Crex Crex Collective, 2018, p. 97)

This touchstone begins with a radical premise—that much of current educational practice, particularly that which rests on the same theoretical footing as modern western culture, is anti-environmental. And that by maintaining the *status quo,* or simply trying to tinker with the edges of what currently exists in modern public education will not be enough to change the human/nature relationship or even limit the destruction being wrought today. As such, Wild Pedagogies is a project of cultural change. This makes educating an explicitly political act and places the teacher in the role of activist recognising that the choices being made in the classroom have explicit and implicit implications for how learners come to understand themselves, what it means to be human, and the importance of the more-than-human world therein.

With this discussion as background, educators might want to consider questions such as:

- Where are my old habits limiting possibilities—in response to existing curricula and values embedded in my workplace? What would it take to be satisfied with my response?
- What am I doing to help learners develop political agency? Am I offering them realistic tools to imagine alternative futures and support change?
- What politics of the natural world have we encountered, and how have we brought them into our learning spaces?
- Have learners been given opportunities to consider their current relationships with places they inhabit and the other beings they live alongside? And, did they have the right to change them?

Touchstone #6: Building Alliances and the Human Community

> We believe that the colonial ethos of resource extraction is not separate from, but is yet another shade of the many hierarchies of dominance that exist amongst humans. For this reason wild pedagogues seek alliances and build community with others not only in the environmental world but across all people and groups concerned with justice. (Crex Crex Collective, 2018, p. 102)

This touchstone seeks to remember the importance of, and to work towards building strong alliances and flourishing communities while at the same time reminding us not to forget the human in all of this work. Here the implicit goal is to push against the challenges of individualisation and alienation and, at the same, to resist the colonial move to separate marginalised groups be they human or other-than and place them at odds with each other. In order to create flourishing equitable communities, we will need to listen and learn from each other. Here educators across the formal and informal spectrums have the opportunity to work with and learn from myriad others. Environmental educators from critical race theorists, community organisers from experiential educators, popular educators from gender theorists, and beyond.

With this discussion as background, educators might want to consider questions such as:

- Who makes up my communities when I think of doing Wild Pedagogies work? Why? Who is left out, but should be included? Why?

- How do I support my communities and how do they support me? How can I foster these same questions in my classrooms or other learning spaces?
- How do my various communities make decisions? Who is affected by these decisions? And are those affected part of the decision-making process?
- What can I do to bring the natural world more explicitly into community decision-making?
- How may my communities encourage one another to depart from the status quo? How do we encourage and challenge each other, allow for mistakes, and rebound from setbacks?

Touchstone #7: Learning That Is Loving, Caring and Compassionate

> We believe that humans are able, if given the opportunity, to develop rich relationships with myriad members of the more-than-human world. And, that these relationships of reciprocal care are part of overcoming the alienation that exists between many humans and the natural world (Jickling & Blenkinsop, 2020, p. 126).

David Orr (2017), like many others, calls for serious educational change, because "without exaggeration it will come down to whether students come through their formal schooling as more clever vandals of the Earth and of each other" on one hand, "or as loving, caring, compassionate, and competent healers, restorers, builders, and midwives to a decent, durable, and beautiful future" (pp. ix–x) on the other. What will it take to nurture caring, compassionate, and competent restorers of the earth?

With this discussion as background, educators might want to consider questions such as:

- What did I do today that required learners to be sensually present in their learning? To encounter the other, to feel, care, respect, and to notice the more-than-human world?
- What have I done to accommodate experiences that exist beyond the capacity of language to fully describe and evaluate?

- Even though some learning cannot be measured, it still exists. How can I create a positive space in my evaluation scheme to honour this existence?
- Have I considered how to hold space for learners as they encounter the range of emotions that appear in response to burgeoning care? What kinds of skills and supports can I offer them as they act in ways that are at times contrary and potentially threatening to the system in which they live and learn?

Touchstone #8: Expanding the Imagination

> We believe that the ecological world has changed dramatically, and that public education must change in response. Future teachers can no longer be trained for a system that leaves students ill-prepared to respond to current crises and imaginatively unable to create new responses. (Jickling & Blenkinsop, 2020, p. 131).

Current ecological precarity requires us to think differently, be different, and imagine differently. We can no longer rely on our own limited problem-solving or creativity—we must find ways of extending our own imaginative thinking beyond ourselves, to draw in the broader more-than-human world. Such thinking requires ecological forms of imagination that reach beyond our own skulls to engage external materials, processes, entities, and places (Morse, 2021). We must be aware that the imagination is not as broad and flexible as suggested in everyday understanding. For educators, this has important implications. Not only do we have to recognise our culturally limited imaginations, but part of our pedagogical work might be to expand the range of options available to teachers. The languages we learn to speak and the foundational stories we are told shape who we are in the world. But they also limit what we can think and imagine.

With this discussion as background, educators might want to consider questions such as:

- What did I do with my practice today that pushed outside the students' previous experiences and my own imagination?
- What new "stuff", experiences, and stories did I add to the mix? And, how are students taking up, working with, and being changed by these diverse cultural tools?

- Did I notice my proclivity to "not do" the seemingly unusual or limit learners seeking to do the same? Or did I make a considered attempt to provide space for the unusual to happen?
- What cognitive, physical, cultural, and natural tools am I working with right now? And what new ones might I try introducing? Where might I look to find additional ideas?
- What are the edges of my experience that might limit how far I can imagine into a different kind of education? What are the limits of my own imagination?
- What sources of inspiration am I seeking to support and enhance pedagogical change and development?

References

Crex Crex Collective: Hebrides, I., Affifi, R., Blenkinsop, S., Gelter, H., Gilbert, D., Gilbert, J., Irwin, R., Jensen, A., Jickling, B., Knowlton Cockett, P., Morse, M., Sitka-Sage, M., Sterling, S., Timmerman, N., & Welz, A. (2018). Six touchstones for wild pedagogies in practice. In B. Jickling, S. Blenkinsop, N. Timmerman, & M. D. Sitka-Sage (Eds.), *Wild pedagogies: Touchstones for re-negotiating education and the environment in the Anthropocene*. Palgrave Macmillan.

Jickling, B., & Blenkinsop, S. (2020). Wilding teacher education: Responding to the cries of nature. *Canadian Journal of Environmental Education 23*(1): 121–138.

Morse, M., Jickling, J., Blenkinsop, S., & Morse, P. (2021). Wild Pedagogies. In G. Thomas, J. Dyment, & H. Prince (Eds.), *Outdoor environmental education in higher education: International perspectives*. New York: Springer. https://doi.org/10.1007/978-3-030-75980-3_10

Morse, P. (2021). Flowing magma bodies: Towards a relational understanding of imaginative pedagogical possibilities. *Environmental Education Research 27*(8), 1229–1244. https://doi.org/10.1080/13504622.2021.1926432

Orr, D. (2017). Foreword. In B. Jickling & S. Sterling (Eds.), *Post-sustainability and environmental education: Remaking education for the future* (pp. vii–xi). Palgrave Macmillan.

Index

K. Malone et al., *Wilding Ecologies, Walking-with Glacier*,
https://doi.org/10.1007/978-3-031-63182-5

SPRINGER NATURE

GPSR Compliance

The European Union's (EU) General Product Safety Regulation (GPSR) is a set of rules that requires consumer products to be safe and our obligations to ensure this.

If you have any concerns about our products, you can contact us on ProductSafety@springernature.com

In case Publisher is established outside the EU, the EU authorized representative is:

Springer Nature Customer Service Center GmbH
Europaplatz 3
69115 Heidelberg, Germany

The manufacturer's authorised representative in the EU is Springer Nature Customer Service Centre GmbH, Europaplatz 3, 69115 Heidelberg, Germany. If you have any concerns regarding our products, please contact ProductSafety@springernature.com

Printed and bound by CPI Group (UK) Ltd, Croydon, CR0 4YY
01/10/2026
02225983-0006